The ethnographer with the headman's son, enjoying the cool waters of the Kaveri. (Rampura 1948)

THE DOMINANT CASTE
AND OTHER ESSAYS

M. N. SRINIVAS

DELHI

OXFORD UNIVERSITY PRESS

BOMBAY CALCUTTA MADRAS

1994

Oxford University Press, Walton Street, Oxford OX2 6DP

Oxford New York Toronto
Delhi Bombay Calcutta Madras Karachi
Kuala Lumpur Singapore Hong Kong Tokyo
Nairobi Dar es Salaam Cape Town
Melbourne Auckland Madrid

and associates in
Berlin Ibadan

Oxford India Paperbacks 1994
Revised & enlarged edition

ISBN 0 19 563465 9

Typeset and printed in India
by Pramodh P. Kapur, Raj Bandhu Industrial Co. New Delhi 110064
and published by Neil O'Brien, Oxford University Press
YMCA Library Building, Jai Singh Road, New Delhi 110001

arbitrator, representing his elderly and powerful father. As the disputes went on, my young friend or one of his brothers would trot out a few facts about them and ask me to give my verdict. On a few occasions, I did rise to the bait and said the first thing that came to my mind. My young friends had no difficulty in showing how unsound, if not downright silly, my verdict was. In order to do this they brought out new facts which had not been told me before. I was shocked at their lack of any sense of fairplay, but it gradually dawned on me that as far as they were concerned, they were only out to get a laugh at my expense. But that was not all. The disputants, their friends, and sometimes, onlookers, began to take notice of what was happening in the arbitrators' corner. They also joined in the laughter at my expense. In a word, the attention shifted suddenly from the disputants and the dispute to me, the observer. I became embarassed at once, and even upset, for somewhere inside I felt that I had violated the privacy and self-respect of the disputants. Much later I realized that I had ruined the very thing that I wanted to observe.

Most disputes had a dramatic dimension which made them stand out in one's memory. A huge haystack was burnt to avenge an old wrong; two men fought each other in the middle of the street drawing blood; a middle-aged Muslim housewife held an audience of men of all ages spellbound with the presentation of her case against her husband's younger brother's wife; and so on. There was a heightening of excitement, and increased interaction among the disputants, their kith and kin, and their friends and enemies, for the duration of the dispute. Disputes afforded opportunities to watch the behaviour of villagers under conditions of stress, sometimes severe stress. Not only that: rules, norms and values were interpreted afresh to meet the exigencies of the dispute. In the process the norms etc., acquired a resilience which enabled them, over a period of time, to be applied to the complex facts of a vast range of disputes.

'Village Studies, Participant Observation and Social Science Research in India', is, as its title suggests, an attempt to explore the implications of the method of participant observation, which was used for the study of Rampura village for enhancing our knowledge of Indian culture and society as a whole, and for social science research. I shall expand on this theme later.

The first essay in the book, 'The Indian Village: Myth and Reality', grew out of an effort to combat the erroneous view so magisterially propounded by Dumont and Pocock, in the very first issue of their influential journal, *Contributions to Indian Sociology*, that the existence of caste differences prevent villages from becoming communities. The basis of this thesis is an unstated assumption that egalitarianism among residents is a prerequisite of community formation. Another implicit assumption of their essay is that such (egalitarian) communities exist in western Europe, no mention being made of the existence of social classes. A total rebuttal of the Dumont–Pocock thesis needed going into unavoidable detail. In fairness to Dumont it must be mentioned that he has characterized this critical essay as a 'well-balanced synthesis' (1980 , Xli).

As mentioned earlier, the concept of 'dominant caste' was first adumbrated in the essay, 'The Social System of a Mysore Village' (1955a), and later elaborated in 'The Dominant Caste in Rampura'. The concept became popular, and is now widely used by those who wish to analyse and comment on power structures in rural India. Initially, I defined a dominant caste in the following terms: 'A caste may be said to be "dominant" when it preponderates numerically over the other castes, and when it also wields preponderant economic and political power. A large and powerful caste group can more easily be dominant if its position in the local caste hierarchy is not too low' (1955a, p. 18). Dumont characterizes this definition as 'rather vague' and 'which must be discussed and made more precise' (1980, p. 161). It may be explained here that the 'vagueness' was deliberate as it was an initial formulation and I intended to take up the idea for a more careful consideration later. Dumont thinks that ownership of land, or the possession of superior rights in land, is the sole source of dominance, and he further argues that the question of the strength of numbers of a caste at the village or other local level is irrelevant to dominance, since landowners are able to obtain the services of the landless through the institution of clientship.

Let me now consider the question whether strength of numbers is really irrelevant to 'dominance'. Dumont himself makes a significant observation in this connection which deserves to be quoted: 'One fact to be stated is that usually, when sufficient data are available, it can be seen that the most numerous castes in a village are, first, the dominant caste, and secondly the caste which provides the greatest part of the labour force and is usually untouchable, rather as if the castes in the closest relation to the land, either in theory and practice, had the greatest possibility of increasing in number. However, this would not be enough to introduce number as one of the criteria of dominance. Srinivas justifies it in an article on the dominant caste in the same village, in which he explains that the real status of one and the same caste in different villages can depend on the number of men it can put into the battle line, and that even Brahmans feel insecure where their numbers are rather small. The emergence of brute force does not cause much surprise. But it does not show that it is necessary for the caste which is powerful in land to be itself numerous, for such a caste easily attracts a clientele' (1980, pp. 161–2).

It is astonishing that Dumont dismisses as of no account the empirical fact (according to him) that both the dominant and untouchable castes—landowning and landless respectively—are represented in some strength at the village level. How can he be so certain that there is nothing underlying the empirical association? Second, it is not enough to say that 'the emergence of brute force does not cause much surprise'. What is the role of 'brute force' in rural society? Apart from the fact that intergroup violence—whether between factions, castes or villages—did occur occasionally, a caste section in a village was able to win the respect of other castes if it had strength of numbers. In the Rampura region, for instance, Fisherman (Besthas) rank much lower than the dominant caste of Okkaligas, but in the villages where they enjoyed numerical strength they were known to stand up for themselves. Okkaliga elders mentioned the existence in some villages of boundary stones marking out the respective territories of Okkaligas and Fishermen. In his facinating book of memoirs, Navaratna Ramarao mentions the successful assertion in the early years of this century by Fishermen in Bannur of their right to take out a wedding procession in the main streets of the village in the teeth of the opposition of the dominant Okkaligas (1954, pp. 203–16). Rampura lies in the Bannur *hobli* division (I substituted the fictitious name of Hogur for Bannur in my earlier writings). Strength of numbers does provide some protection against the worst forms of oppression and abuse, particularly for non-Untouchable castes. In other words, strength of numbers can be translated into social rank. In conversation after conversation, informants referred to particular villages or towns as places where their caste enjoyed strength of numbers. This was a matter of pride for them, and disputes within the caste were taken there for settlement. The opposite of this phenomenon was found among some Brahmins in the Rampura region who occasionally felt that they should move into towns for greater security. In the early fifties the Okkaliga caste headman of Kere was so

annoyed with the Brahmin village accountant family that he seriously contemplated asking them to leave Kere for their own good. As
the accountant told me, the headman said that he would offer them
betel leaves, arecanuts and coconuts on a tray as a gesture of bidding
them farewell. The accountant thought that this only showed that the
headman was woefully out of tune with the times. The accountant
and his relative had worked for the headman's rival, also an Okkaliga,
for the State Assembly seat from the Kere region. The accountant's
patron won and hence the headman's wrath. The patron was influential and would certainly try and protect his client, but he stayed in the
State capital of Bangalore and not in Kere.

Strength of numbers plus a tradition of violence are essential
elements of dominance, and this is integrally linked to the character of
the pre-British political system. It was a system in which warfare was
endemic, particularly at the lower levels of the system. The death of
a ruling raja or chief frequently gave rise to succession squabbles
among the heirs and their supporters, squabbles which not infrequently led to violent conflict and chaos. Such disturbances afforded
opportunities for politically ambitious men from the dominant castes
to seize power at the level of the little kingdom and region. While
clientship was important in binding together members of different
castes together, it alone was not enough to ensure a miniscule caste
owning a great deal of land to become dominant. The only exception
that I can think of is when there is a small nucleus of people whose
claims to royalty are beyond question. In such an event, strength of
numbers would not be crucial.

Dumont has commented on the homology between 'the function of
dominance at the village level and the royal function at the level of a
larger territory' (1980, p. 162). Again, ' . . . the homology extends so far
that the dominant caste is often a royal caste, a caste allied to royal
castes (Mayor), or a caste with similar characteristics (meat diet,
polygyny, etc.)' (1980, p. 163). (The last clause would make 70 per cent
of the Hindu population 'royal'!) Dumont does not ask the question,
what is the source of such homology. In the last analysis, the rights of
ownership which the dominant caste members have in their land
should have the approval of the raja, and as already mentioned,
leaders of dominant castes entertained ambitions of becoming chiefs.
They also wanted their castes to have the status of Kshatriyas in the
various hierarchy. It is no wonder then that the royal model made a
deep impression on the dominant castes. But that, however, was not

the only model available to them. The Brahminical model had its appeal, both directly, and as mediated through the Kshatriyas. It is well to recall here that there are at least three models of Sanskritization, Brahminical, Kshatriya and Vaishya, and the dominant castes seem to have been subject generally to the pulls of the first two. The raja or king conferred both political and economic legitimacy on the dominant caste, while the Brahmin was indispensable for legitimizing higher caste status. If a great temple or pilgrimage centre was located near the habitat of the dominant caste, the Sanskritic-Brahminical pull was likely to be greater than the royal or Kshatriya pull. At the other extreme was a dominant caste which wielded power locally but was far way from the influence of Kshatriyas or Brahmins. Such a situation was frequently found in tribal areas. While this was generally true, over the millinea Hindu religious ideas travelled far and wide into the forested mountains through sanyasis and pilgrims.

To trace our steps backwards for a while: it is hard to understand Dumont's dismissal as of no consequence the empirical fact mentioned by him that both dominant castes and Scheduled Castes (landless labourers) are represented in strength in rural areas. His other argument amounts to saying that landownership is everything and numbers are nothing: the landowning caste is able to control the others resident in the village through the institution of clientship which gives the latter dependent rights in land, or to a share in the produce of a specific piece of land. Dumont's bland assumption that clientship always meant total subservience and loyalty to the patron is difficult to justify. Factions were always a part of the rural social structure and marginal clients were prone to change patrons. Besides, there was always the horizontal tie of caste pulling against the patron-client tie.

It is well to remember in this context that in pre-British India, it was labour that was scarce and not land, and therefore, the services of good agricultural labourers and skilled artisans were always at a premium. Splits in dominant caste lineages, and factions drove individuals to found new villages and this meant the clearance of thick jungle, a task calling for heavy investments of labour. One expects that the labour came largely from the dominant castes and their servants, primarily landless labourers. This type of situation accords well with the kind of empirical situation mentioned by Dumont that castes with close ties to land and agriculture are numerically strong. But in recent years dominance is tending to tilt more and more towards castes which are populous but which do not own a significant quantity of locally available land. This is of course due to adult franchise and an increasing awareness of their rights on the part of the poor everywhere.

The fact that ballots are secret has meant the conferring of greater independence in voting to the client castes. As far back as 1952, the Untouchables of Rampura took advantage of the secret ballot to actually vote for their own candidate while promising the village headman that they would certainly follow his instructions in voting. The defeat of 'his' candidate angered the headman and he abused the Untouchables for having double-crossed him. Some days previously I had heard that the leaders of Untouchables had gone round the villages of the region asking their followers to cast their votes for candidates who would promote the interests of Untouchables and not for those who merely obeyed the leaders of the dominant castes.

I have earlier stated that the absence of numerical strength did not matter where there existed a well established royal lineage, and I think that I should mention another exceptional situation: when a small number of Brahmins owned a large quantity of local land, they could exercise dominance in village affairs. This was due to their ritual pre-eminence. But since independence the forces of democracy and secularization have been so strong that power tends to move inexorably in favour of numbers. But—and here is the crucial fact— since castes which are dominant today are both numerically strong and own a substantial quantity of local land, they are clearly the most powerful section of the rural population in India. The poorer members of the village are bound to them by ties of clientship. Factions weaken dominant castes no doubt but then they also weaken the dependent castes. Weakening clientship makes clients less reliable but the presence of large numbers of the poor and landless ensures a constant supply of clients to rich patrons. *All in all, post-independent India is, certainly at the regional if not at the state level, the India of dominant castes.* The dominant castes are prominent in politics and the professions, and they have left their mark on every institution, and on the culture of each state. This has been accomplished more thoroughly in those states which possess very large and powerful dominant castes such as Andhra Pradesh, Maharashtra and Karnataka. The phenomenon of the omnipresence of the dominant castes is recognized by the people and even journalists, but social scientists seem to turn a blind eye to it.

Dominant castes performed kingly functions in a very real sense at the village, and sometimes, also at slightly higher levels. The informal council of the leading elders of the dominant caste maintained law and order in the village, the justice that they dispensed being simple and prompt, made sense to the people. Another important function of the dominant caste was to maintain the social order, and in particular, to make sure that each caste performed its duties and did not assign to itself the rights and privileges which belonged to the higher castes. Dumont and Pocock link this function of the dominant caste with the fact of ambiguity with regard to its position in the local hierarchy: 'Although they are masters in terms of power, their status is not so absolute that they can allow promiscuity of service' (1957 : 31). This is an insightful observation but while the dominant caste could not allow the artisan and servicing castes to claim any rights or privileges which would endanger the dominant caste's own position in the local hierarchy, it did feel responsible for the maintenance of the social order just as the chief or raja felt a similar responsibility at the higher levels.

Decentralization was an integral feature of the pre-British political system in as much as it was a function of the pre-industrial technology of India. Roads were very poor if not non-existent, and villages were isolated from each other and from towns. Pre-British India, in spite of the fact that vast and sophisticated empires flourished in it, was characterized paradoxically by 'pedestrian states', the state's duty to maintain law and order being confined to towns and to the few highways which existed. Supplementing this, the dominant castes maintained law and order at the local levels.

When a caste, or more properly speaking, its local section, possessed one attribute of dominance, there was a tendency for it to attract to itself the other attributes. This applied particularly to the economic and political elements of dominance: that is, when a caste had political power it was able to attract to itself wealth, particularly in the form of land, and, on the other hand, when it had wealth it was able to attract to itself political power, though here limitations were imposed by the local size of the caste particularly in relation to the size of the other castes, and such other factors as the strength of caste loyalty.

When there was ritual pre-eminence, as in the case of Brahmins, it led traditionally to their acquiring wealth, and even influence, if not power. Giving gifts to Brahmins was regarded as conferring religious merit on the donors, and rajas gave land, house and gold to deserving (i.e., poor, pious and learned) Brahmins. Since Brahmins were the most literate section of the society they were employed by their patrons, kings, chiefs and rich landowners, in a variety of jobs, as civil servants, clerks, accountants, legal experts, doctors, astrologers, and the like. When the ritual eminence of the Brahmin was combined with landed wealth and bureaucratic position, he was a force to reckon with. However, the democratic, secular and egalitarian winds which are blowing in the country today have resulted in an erosion of the position of the Brahmin. Many intellectuals among non-Brahmins believe the caste system to be a diabolical creation of the Brahmins to keep themselves perpetually at the top and the others at the bottom. The theory was politically very handy in that it showed the proponents of the theory in a favourable light, as opposed to all inequality and exploitation in all their myriad forms. Even more, it disguised the fact that the dominant castes, who are generally non-Brahmin and owners of the land, are among the worst practitioners of inequality and exploitation, their victims being the local poor, particularly the landless members of the Scheduled Castes.

However, the dominant castes are not the only practitioners of inequality. It occurs at lower levels inclusive of the Scheduled Castes who form a hierarchy among themselves. One of the commonest—and rather cynical—features of the present movement towards equality is that each caste regards itself as the equal of castes superior to it while simultaneously denying similar claims from those inferior to it. The rhetoric that is used by the new egalitarians frequently hides this unpleasant truth.

Dumont objects strongly to my use of the expression of 'ritual dominance' (1980, p. 1627), since in Dumont's formulation status or religious rank is the polar opposite of power, or dominance. Dumont claims that his formulation is derived from the varna system itself. He writes, 'We prefer to maintain a fundamental distinction between the two (status and dominance), a distinction which, as we have seen, is built into the theory of the varnas itself, if not into that of the castes' (1980, p. 162).

My use of the term 'ritual dominance' was derived from the empirical fact that in purely ritual or religious contexts the Brahmin did enjoy a certain superiority to his wealthy patron, a superiority that appeared close to power. I have watched this phenomenon even among different kinds of Brahmins. The secular, educated and well-off Brahmin who employs a poor Brahmin priest to perform a ritual may be regarded as enjoying a general kind of superiority to the priest. But during the ritual, even when there are arguments about the fees to be paid, the priest enjoys a contextual superiority to his patron, a superiority derived from his knowledge of the ritual and his indispensability. A classic example of this kind of positional shift I saw between the priest of the Rama temple and the headman of Rampura. The secular—economic and political—dependence of the Rama priest sharply contrasted not only with his superiority in ritual contexts but his ability to influence the behaviour of the headman for the duration of the ritual. The priest not only had a higher ritual rank than the headman but he was the representative of a religious and theological system which commanded the respect and allegiance of all, including the headman. It was that respect and allegiance which made the headman a force for conservatism in the village.

One of the consequences of dominance is that landowners from the dominant caste became patrons to clients from the artisan, servicing and labouring castes. Within the village, the patron—client relation manifested itself mainly in the *jajmani* system. However, not all patrons were of the same level : the majority of them were small, and often, themselves clients of bigger patrons, while at the top were the leaders of the village factions. It may be remarked here that factionalism was integral to the twin phenomena of dominance and patronage. Factionalism seems to be as much a part of the social structure as caste, though there is an odd reluctance on the part of social scientists to recognize it as such. Instead, factionalism is looked upon as a kind of disease, and a curable one. While caste provides a horizontal type of bonding, faction provides a contrary type of bonding, one that is vertical, multi-caste, and cross-class. The two types of bonding are simultaneously opposed and complementary, and integral to the structure.

Peter Gardner makes an interesting distinction between 'patronhood' and Kshatriyahood: 'In earlier sections of this paper, I have attempted to show that patronhood constitutes a simple and direct form of dominance. Kshatriyahood, by contrast, is much more a matter of superstructure than foundation. Nevertheless, the claim to Kshatriyahood is pan-Indian, it has broad structural implications that cannot be passed over lightly' (1968, p. 89).

Peter Gardner and S. C. Dube concede that while numerical strength is a *necessary* condition of dominance, it is not a *sufficient* condition. Gardner states, for instance, ' . . . it is not so clear that numerical preponderance is sufficient to *create* a dominant caste unless used as a means to wrest ownership of the village land' (1968 :

86). Dube makes more or less the same point when he remarks, 'Numerical strength, while it is an element of dominance, does not necessarily make a caste dominant' (1968, p. 59).

Dube has some serious objections to the concept of dominant caste : 'It will be meaningful to speak of a 'dominant caste' only when power is diffused in the group and is exercised in the interest of the whole group or at least a sizeable part of it. When there are pronounced inequalities of wealth, prestige and power between different individuals in a so-called dominant caste, and where dominant individuals exploit the weaker elements in their own caste as well as the non-dominant castes it will perhaps be inappropriate to think of it as a dominant caste. Unity and concerted action in terms of caste interest, therefore, must be assumed before we locate dominance in a caste on the basis of criteria specified by Srinivas. The fact that a number of dominant individuals occupying most of the power positions belong to a particular caste is, by itself, not enough to characterize the caste as dominant' (1968, p. 59).

Dube also makes dominance of a caste dependent upon the existence of unity within it: 'Intra-caste unity and articulation in terms of power are essential for its emergence as a dominant caste. Where these conditions exist a caste can become dominant. In their absence, however, the community power structure can be better understood in reference to dominant individuals, dominant faction and their complex alignment' (1968, p. 61).

Dube concedes the existence of 'dominant individuals' and 'dominant faction' but he is hesitant to go the whole hog and concedes the existence of 'dominant caste'. But it is our basic contention that both 'dominant individuals' and 'dominant faction' owe their dominance to the fact that they are a part of the dominant caste. It may be observed here that most, if not all, the leaders of the 'dominant faction' hail from the dominant caste except in areas where there are two rival castes each striving to establish its dominance. Again, the unity of a caste, particularly that of a dominant caste, is not something static and constant, but dynamic and contextual. It emerges especially in a relation of opposition to the other castes.

It is a well-known fact that caste is used for mobilizing political support, and this was much greater in the years immediately following independence than now. Even now, at the lower levels of the political system viz., state, region, district, tehsil and village, caste loyalties are freely tapped for winning votes. Obviously, caste is not the only factor. Also, the fact that the term for caste, *jati*, has a semantic stretch which enables cognate sub-castes to come together during elections is helpful for the political mobilization of castes. And as has already been mentioned, patrons from the dominant caste are able to obtain the votes of dependent castes thanks to the institution of clientship though this is becoming increasingly difficult. The poor and lower castes are more and more inclined to vote in their own interests, instead of voting in the interests of their patrons. This is associated with the growing tendency for political mobilization to follow the lines of caste.

Dube is really confusing the issue when he remarks that it is meaningful to speak of a dominant caste 'only when power is diffused in the group and is exercised in the interest of the whole group or at least a sizeable part of it'. I must make it clear in this connection that I was concerned with caste being used as a base for acquiring power and *not* with questions of how that power was being used by the leaders of the dominant caste after they had acquired it. Nor for that matter, whether there was equality among the different members of the dominant caste. These are, to say the least, irrelevant, even if we ignore the problems that exist in deciding the question whether power was or was not being used in the interests of the caste as a whole.

But, surprisingly, in a conclusion that is at variance with his earlier argument, Dube concedes the entire case for the concept of the dominant caste: 'The process of group dynamics and the web of power alignments in these (four) villages are infinitely complex. For understanding them the dominant caste frame of reference, by itself, does not appear to be adequate. Without doubt persons from some of the upper castes have a larger proportion of land and literate persons. They also have the added advantages of relatively higher ritual status. This opens for them the avenues to power positions in the village and persons from these categories do in fact occupy most of the top positions. But it should be borne in mind that they derive a part of their strength from caste; support of other castes is no less important. Caste is not the power-wielding unit, multi-caste power alliances are' (1968 : 80). I have never claimed that dominant caste provided the *total* explanation for the phenomenon of power in rural India. As for multi-caste alliances, Dube has forgotten the fact that dominance expresses itself in patronship, and patronship is inconceivable without clients.[1]

---

[1] See in this connection T. K. Oommen 'The Concept of Dominant Caste : Some Queries', *Contributions to Indian Sociology*, NS, IV (1970), 73–83. Oommen makes more or less the same points as Dube (1968), and hence a separate reply was not thought necessary.

'Village Studies, Participant Observation and Social Science Research in India', grew directly out of my fieldwork in Rampura. Research using the method of participant observation was practically unknown in the country when I carried out my field-study of Rampura in 1948. It was after my return to India in June 1951 to take up the professorship of sociology in the M.S. University of Baroda that I began to advocate the method of participant observation in the study of tribal and rural communities, and other isolable groups. I think my students at Baroda, and later at Delhi, were among the first indigenously-trained scholars to carry out intensive fieldwork in the country.

During the years immediately following India's independence, several social anthropologists and sociologists from abroad, in particular, the USA, and UK, carried out intensive field-research in India's villages and in tribal and urban areas. But while the research was intensive, involving staying in, or near the field, very few did research by using the method of participant observation in the Malinowskian sense, dispensing with the use of interpreters and assistants, and communicating with the indigenes in their own language. But one of the results of the considerable research activity in the '50s and '60s was to make it clear to Indian social scientists that the survey method was not the only one available for carrying out empirical research in the social sciences.

During the '50s a series of articles appeared in the *Economic Weekly* (as it was then known) of Bombay, giving brief accounts, in plain English, of villages, and tribal communities, in which anthropologists, Indian, American and English, were carrying out intensive fieldwork. Fifteen such contributions were published in all and they were included in a book entitled *India's Villages* (1955b). This modest effort had an unexpectedly warm welcome from social scientists, planners, officials and others. Following independence, there was a sharp rise of interest in rural India and its development under the auspices of centralized planning. At a more professional level, *Village India*, edited by McKim Marriott (1955), *Aspects of Caste in South India, Ceylon and North-West Pakistan*, edited by E. R. Leach (1960), and S. C. Dube's *Indian Village* (1955), all served to popularize village studies, and incidentally, make it clear to non-anthropologists that anthropologists had a distinctive contribution to make to the understanding of Indian rural society and culture. It would not be an exaggeration to state that anthropology brought in a breath of fresh air to the hitherto arid area of village studies characterized by over-reliance on quantitative methods and in which very little information was available about the culture and institutions of villagers.

The dominant style of social science research in India has been the survey method in which the actual collection of data is through the 'canvassing' of questionnaires among selected samples of human beings. A problem which has policy implications, or which is thought to have policy implications, is investigated through the survey method, and integral to this method, at least as it is practised in India, is the division of labour between the designers of the study and the analysts of data on the one hand, and the people who actually collect the data on the other. The latter are again hierarchized into different grades, from supervisors to investigators.

There are innumerable agencies in the country collecting information on a variety of matters, and India must be one of the most assiduous countries in this respect. Indeed, what is now needed is an annotated directory or compendium of data-gathering agencies, governmental and non-governmental and at various levels from the national to the local. It would be very useful to find out how much of this data is utilized for purposes of the formulation of policies and the implementation of programmes. A cynic might feel that the gathering of social data has become an end in itself in India: ministers and officials seem to support the funding of research projects in the social sciences more to prove their own sophistication than because they are necessary to achieve administrative and social goals. The reports that the researchers write are generally allowed to gather dust on the shelves till one day some Ph. D. student stumbles on them and brings their findings to light.

The money spent by various agencies conducting survey research is considerable, especially for a poor country like India, and in contrast, the amount spent on intensive studies is pitiful. Until recently intensive studies of communities were regarded as a wasteful eccentricity of a few scholars who could not do anything better. The main opposition to the intensive studies of the anthropologists came expectedly from those who had carried out survey research and who regarded the intensive study of small communities as mere exercises in academic curiosity and incapable of assisting policy formulation or programme implementation. In the august bodies which decide on the projects to be funded, the term, 'of academic interest only', is frequently used pejoratively.

In recent years, however, there have been efforts by the practitioners of the two distinct approaches to try and learn from each other,
and to try and see if the two can not be made mutually complementary. There are several reasons for such coming together and
one of them is the widely-felt conviction that India is such a vast and
diverse country that if development has to spread to all parts of it an
intimate knowledge of the culture, economy and social life of each
region is essential. This conviction has gained strength from the fact
that even though development is occurring in the country as a whole,
it is unevenly distributed, and further, the benefits intended for the
poor do not often reach them. This has naturally led to an interest in
the processes which prevent the well-intended measures of the
government (and of even voluntary agencies) from reaching their
targets.

However, a distinction needs to be made between micro-level
research and research using the method of participant observation.
For micro-level research can be—and in very many places in India,
is—carried out using the routine survey methods whereas participant
observation has been used so far only in the study of small groups and
communities. It is basically a qualitative method where the anthropologist learns about the culture and institutions of the people he is
studying through living with them for a year or two, speaking their
language, and sharing their experiences and living conditions. In
participant observation, it is not enough to gather information, but
necessary to try and obtain an understanding of the people one is
studying. Such understanding involves the total psyche of the anthropologist and not merely his intellect.

This is not the place for a full review of the pluses and minuses of the
two seemingly opposed approaches to the understanding of social
science problems, but a few familiar criticisms of the method of
participant observation may be mentioned here. There is a widespread feeling among Indian social scientists that scientific activity is
synonymous with quantification, and that the greater the use of
sophisticated mathematical and statistical techniques the more scientific the activity. Since participant observation is basically a qualitative method, making minimal use of quantitative techniques, there is
a strong suspicion that it is not science. There is now more or less
unanimity among social anthropologists who have had experience of
using the participant observation method that the personality of the
investigator cannot be separated from his observations and interpre-

tations. This is viewed by many social scientists in India as giving into rank subjectivism, and as far from objective scientific activity as possible. There is another and widespread criticism directed against participant observation: how can one generalize from a single village (or tribal) study? The unstated assumption here is that it is only the generalizing activity of the scientist which makes science valuable. Generalizing, particularly in this context, is linked with policy formulation which, in turn, leads to programmes and their implementation. In short, the pursuit of knowledge which is unrelated to these activities, is sneered at.

The criticism that participant observation is time-consuming also comes generally from those—and they are many—who believe that the true goal of scientific activity is to help governments in formulating correct policies and devise suitable programmes in pursuit of those policies. From this point of view, participant observation, aimed as it is at obtaining accurate information about, and an understanding of, the cultural and social life of particular communities of tiny regions, might seem a luxury particularly for Third World scholars. This, again, is the result of taking too narrow a view of the method: participant observation often uncovers hidden relationships between seemingly discrete areas of social life and these can occasionally be translated into hypotheses for further testing. Such relationships often elude investigators in a hurry. Besides, one needs to look at cultures and social systems as wholes in order to be able to detect hidden or far-out relationships. And participant observation is the method which enables the anthropologist to detect such relationships.

As to subjectivity, it is there in all social science observation though everyone may not be aware of it, or in the same measure, and in fact, it is only in so far as the social scientist is aware of his subjectivity and its contours, that he moves towards greater objectivity in his work.

While participant observation is not aimed at producing generalizations, in studying a unit of the total social system and culture, the anthropologist is looking at the microcosm which in a way reflects the macrocosm, and an intensive study of the former does provide insights into the latter. If I may refer to my own experience, my study of the Coorgs produced the concept of Sanskritization while my study of Rampura produced the concepts of 'dominant caste' and 'vote banks'. Now all these concepts are being used freely in the analysis of cultural and social processes everywhere in India if not south Asia as a whole. In short, while studies of particular communities will not yield generalizations which hold good for the country as a whole, insights into the nature of the macrocosm, are a different matter. Further, such insights can be translated into hypotheses for testing.

Insights into the nature of the macrocosm are particularly helpful in reading historical accounts of the region. The anthropologist can extrapolate from the present into the past, in fact, read history backwards, and thus obtain more insight into the nature of historical processes. Of course, in this enterprise of reconstructing the past, imagination should be tempered by abundant caution.

In short, the intensive studies of small communities ought to be viewed as an essential supplement to macro-surveys. The latter will provide correlations between discrete phenomena while the former are likely to reveal the events and processes underlying the correlations. The scholar who conducts an intensive study into a small community should have, at the end of his analysis, some understanding of the nature of the linkages between different aspects of the society, linkages which are not visible on the surface. But the question of the existence of such linkages over a region can only be ascertained by empirical inquiry.

Macro-surveys also provide prespectives for intensive studies. A macro-survey ought to be able to provide some idea of the differences between cultural regions. For instance, students of caste know that the roles and ranks of Barbers and Washermen are significantly different in north and south India. (This is not to deny the existence of further differentiation in each region.) Such knowledge enables us to locate distinct culture areas within the country as also the transition zones between one culture area and another. But, then, the culture area for one major feature or trait may not overlap with the culture area for another. Some clustering of tracts is, however, necessary to constitute a region.

In a word, then, marco-surveys and micro-studies are mutually complementary, and each could—and should—be used to enrich the other.

REFERENCES

DUBE, S. C., 1955, *Indian Village* (Cornell University Press, Ithaca, New York).
————, 1968, 'Caste Dominance and Factionalism', *Contributions to Indian Sociology*, NS, II : 58–81.

DUMONT, L., 1980, *Homo Hierarchicus* (University of Chicago Press, Chicago).

—— and D. F. POCOCK, 1957, 'Village Studies', *Contributions to Indian Sociology*, I : 23: 41.

GARDNER, PETER M., 1968, 'Dominance in India: A Reappraisal', *Contributions to Indian Sociology*, NS, II : 82–97.

LEACH, E. R. (ed.), 1960, *Aspects of Caste in South India, Ceylon and North-West, Pakistan* (Cambridge University Press, Cambridge).

MARRIOTT, McMKIM (ed.), 1955, *Village India* (Chicago University Press, Chicago).

OOMMEN, T. K., 1970, 'The Concept of Dominant Caste: Some Queries', *Contributions to Indian Sociology*, NS, IV : 73–83.

RAMARAO, NAVARATNA, 1954, *Kelavu Nenapugalu* (in Kannada) (Jeevana Karyalaya, Bangalore).

SRINIVAS, M. N., 1955a, 'The Social System of a Mysore Village', in Marriott (1955).

—— (ed.), 1955b, *India's Villages* (Government of West Bengal, Calcutta).

——, 1976, *The Remembered Village* (Oxford University Press, Delhi).

The Indian Village : Myth and Reality[1]

I

Since the beginning of the nineteenth century the Indian village has been the subject of discussion by British administrators, scholars in diverse fields, and Indian nationalists. The early administrators' reports, a few of which were included in documents placed before the British Parliament, obtained wide circulation owing to the fortuitous circumstance that two outstanding thinkers of the nineteenth century, Karl Marx and Sir Henry Maine, made use of them in the course of their reconstructions of the early history, if not prehistory, of social institutions the world over. Both the administrators' reports and the writings of Marx and Maine influenced the thinking of Indian nationalists and scholars. The first-hand and intensive studies of villages carried out by social anthropologists since the end of the Second World War have necessarily resulted in a critical appraisal of the earlier views and conceptions. Since some social anthropologists had themselves been influenced, consciously and unconsciously, by the earlier views their critical examination of them may be regarded as an attempt at self-exorcism.

Anthropologists who are active in researching into village India are subjecting to critical examination not only the views of earlier writers but also those of their colleagues. Thus, in 1957, Dumont and Pocock asked the question whether the village was indeed the '*social fact*' which it has for so long been assumed to be' (emphasis mine; Dumont and Pocock, 1957, p. 23). Again, 'A field worker takes a village as a convenient centre for his investigations and all too easily comes to confer upon that village a kind of sociological reality which it does not possess' (p. 26). They conclude that 'the conception of "village solidarity" which is said to "affirm itself" seems all too often to be a

[1] The writing of this paper was made possible by a fellowship at the Center for Advanced Study in the Behavioural Sciences at Stanford. I must thank Miss Miriam Gallaher, Research Assistant at the Center, for editorial help in the preparation of this paper.

presupposition imposed upon the facts' (p. 27). 'Village solidarity is nothing other than the solidarity of the local section of the dominant caste, and the members of the other castes are loyal not to the village as such but to the dominant caste which wields political and economic power' (pp. 27–9).

In contrast to the village, caste has 'social reality'. The village is only the dwelling-place of diverse and unequal castes.

> Inequality, which the first [British] administrators did not stress because they found it natural and inevitable, disappears from the picture for many modern Indians, who assume a 'community' to be an equalitarian institution. In contrast to this widespread mentality are the outright statements of Percival Spear, of O' Malley, and of Srinivas: 'In a joint village, there are two classes of men, one with proprietary rights, and the other without them, power resting exclusively with the former'. (Dumont, 1966, pp. 75–6 n.)

The first influential account of the Indian village appeared in the celebrated *Fifth Report from the Select Committee on the Affairs of the East Indian Cy.* (1812), and Louis Dumont has traced its authorship to one of the great British administrators, Sir Thomas Munro (Dumont, 1966, pp. 70–1). After listing the various village functionaries the *Fifth Report* concluded:

> Under this simple form of municipal government, the inhabitants of the country have lived from time immemorial. The boundaries of the village have been but seldom altered; and though the villages themselves have been sometimes injured and even desolated by war, famine and disease, the same name, the same limits, the same interests and even the same families, have continued for ages. The inhabitants gave themselves no trouble about the breaking up and divisions of kingdoms; while the village remains entire, they care not to what power it is transferred or to what sovereign it devolves, its internal economy remains unchanged. (*Fifth Report*, 1812, pp. 84–5.)

The above statement represented both an oversimplified and idealized account of the village in pre-British India and with slight alterations it was to be repeated by writer after writer for the next 150 years. The next influential account of the Indian village was in Sir Charles Metcalfe's Minute included in the *Report of the Select Committee of the House of Commons*, 1832 (Vol. iii, Appendix 84, p. 331). Metcalfe revived Munro's characterization of villages as 'little republics' which were 'almost independent of foreign relations'. Instead of the unchanging internal economy of the *Fifth Report*, Metcalfe used the expression, 'having nearly everything that they want within themselves'. (It is surprising to find even Dumont crediting the Indian village with 'economically almost perfect self-sufficiency'.) Finally, the village communities to which their inhabitants had a profound attachment as evidenced by their returning to them after periods of war, famine, and pestilence, were responsible for the preservation of the people of India, their freedom and happiness.

According to Dumont, 'For the observer of things Indian, there is something idyllic and Utopian about them [the descriptions], and a reader of Stockes' [*sic*] admirable book [1959] is tempted to father this idealization on the romantic and paternalist minds of the period: Munro, Elphinstone, Malcolm and Metcalfe' (Dumont, 1966, pp. 68–9). Romantic and paternalist they probably were, but how could they have failed to take note of the grinding poverty, disease, ignorance, misery, and inequalities of day-to-day living in the villages?

Both Marx and Maine made their own contribution to extant oversimplifications and misconceptions about the nature of Indian villages. According to Dumont, 'Athough Marx and Maine are poles apart in other respects, they come together retrospectively as the two foremost writers who have drawn the Indian village community into the circle of world history. In keeping with contemporary—Victorian—evolutionary ideas and preoccupations, both saw in it a remnant or survival from what Maine called "the infancy of society" ' (1966, p. 80). Both saw in nineteenth-century India the past of European society. Yet another point of agreement between the two was their belief in the absence of private ownership of land in India.[2] When pristine communal ownership of land was viewed alongside political autonomy, economic autarky, and vast numbers of people living in tiny republics which lasted while dynasties above them rose

[2] See in this connection Daniel Thorner's important paper, 'Marx on India and the Asiatic Mode of Production', *Contribution to Indian Sociology*, 9 (1966), 33–6. 'The key to understanding the history of countries like Persia, Turkey and India, he and Engels decided, was the absence of private property in land' (p. 42). 'Without private ownership of land and resulting class antagonism, Asia had never started on the road to development' (p. 43). But Marx seems to have had occasional doubts about the absence of private party in land: 'At other times, Marx was less sure that there had never been private property or at least private possession of land in India. He wrote to Engels that among the English writers on India the question of property was a highly disputed one. In the broken hill country south of the Krishna river, ' . . . property in land does seem to have existed . . . . In any case it seems to have been the Mohammedans who first established the principle of "no property in land" in Asia as a whole.' In some of the small Indian communities, Marx noted in the same letter, the village lands are cultivated in common. In most cases, however, 'each occupant tills its [*sic*] own field.' 'The waste lands were used for common pasture' (p. 43).

and fell, there emerged a picture of a happy pre-colonial past which educated Indians found nostalgic. Pristine communal ownership was interpreted to mean the absence of economic inequalities. The destruction of Indian handicrafts, especially handlooms, for which India was famous in the pre-nineteenth-century world, owing to their inability to compete with goods produced by the British factories and mills, provided a potent case for ending alien rule which had brought in so much misery.

Marx, who was totally preoccupied with economic and social change, found Indian villages singularly resistant to change. He made no attempt to conceal his dislike of them and all that they implied and stood for:

> ... these idyllic village communities ... had always been the solid foundation of oriental despotism ... they restrained the human mind within the smallest possible compass, making it the unresisting tool of superstition, enslaving it beneath traditional rules, depriving it of all grandeur and historical energies. We must not forget the barbarian egotism which, concentrating on some miserable patch of land, had quietly witnessed the ruin of empires, the perpetuation of unspeakable cruelties, the massacre of the population of large towns, with no other consideration bestowed upon them than that of natural events, itself the helpless prey of any aggressor who deigned to notice it at all. ... We must not forget that these little communities were contaminated by distinctions of caste and by slavery, that they subjected man to external circumstances instead of elevating man [to be] the sovereign of circumstances, that they transformed a self-developing social State into never-changing natural destiny. (Thorner, 1966, p. 41.)

All this explained the static nature of Indian society and its passivity, and British rule, while exploitative, had set in motion economic forces leading to the welcome destruction of traditional Indian society. Britain was producing 'the only social revolution ever heard of in Asia. [Marx] believed that England had a double mission in India: to annihilate the old Asiatic society and to lay the foundations of a Western society' (Thorner, 1966 , p. 42).

According to Marx, the basis of the self-sufficiency of the village was in the 'domestic union of agricultural and manufacturing pursuits'. 'The "peculiar combination of hand-weaving, hand-spinning and hand-tilling agriculture" gives the vilages self-sufficiency. The spinning and weaving are done by the wives and daughters' (Thorner, 1966, pp. 38–9). Again, 'the simplicity of the organization for production in these self-sufficing communities' provided the key to the secret of their immutability 'in such striking contrast with the constant dissolution and refounding of Asiatic States, and the never-ceasing changes of dynasty. The structure of the economic elements of society remains untouched by the storm-clouds in the political sky' (Thorner, 1966, p. 57).

Marx also noted the existence of caste and slavery in the context of the village. He referred to the prevalence of a strict division of labour within the village which operated 'with the irresistible authority of a law of nature' (Thorner, 1966, p. 56). But he was not able to weave slavery into his analysis of the village presumably because in his ideal scheme it was a characteristic of a later stage of economic development. It did not really belong there. He just contented himself with mentioning it.

Marx had no kind words for British rule either. He denounced the commercial exploitation of India by the East India Company.

> India, the great workshop of cotton manufacture for the world, since immemorial times, became now inundated with English twists and cotton stuffs. After its own produce had been excluded from England, or only admitted on the most cruel terms, British manufactures were poured into it at a small and merely nominal duty, to the ruin of the native cotton fabrics once so celebrated. (Marx, 1853; Avineri, 1969, p. 106.)

Again,

> It was the British who broke up the Indian hand-loom and destroyed the spinning-wheel. England began with driving the Indian cottons from the European market; it then introduced twist into Hindustan and in the end inundated the very mother country of cotton with cottons. ... British steam and science uprooted, over the whole surface of Hindustan, the union between agriculture and manufacturing industry. (Marx, 1853; Avineri, 1969, pp. 91–2.)

Marx was strengthening the armoury of the Indian nationalists.

The idea of primitive communism of property was a basic idea of the nineteenth-century evolutionists, and Maine saw in the village communities of North-Western Provinces (later called United Provinces) the prevalence of communal ownership. Generalizing from partial information he declared,

> We have so many independent reasons for suspecting that the infancy of law is distinguished by the prevalence of co-ownership, by the intermixture of personal with proprietary rights, and by the confusion of public with

private duties, that we should be justified in deducing many important conclusions from our observation of these proprietary brotherhoods, even if no similarly compounded societies could be selected in any other part of the world. (Maine, 1906, pp. 257–8.)

The 'Village Community' is 'known to be of immense antiquity', great stability, and it is 'more than a brotherhood of relatives and more than an association of partners'. 'It is an organized society, and besides providing for the management of the common fund, it seldom fails to provide, by a complete staff of functionaries, for internal government, for police, for the administration of justice, and for the apportionment of taxes and public duties' (Maine, 1906, pp. 252, 256).

After his experience as the Law Member of the Viceroy's Council (1862–9), Maine came to have a slightly better idea of the complexity of Indian villages: 'Even when the village-communities were allowed to be in some sense the proprietors of the land which they tilled, they proved on careful inspection not to be simple groups, composed of several sections, with conflicting and occasionally with irreconcilable claims' (Dumont, 1966, pp. 83–4).

Dumont has criticized Maine for his failure to understand that 'the constitution of the village had to be put in relation to caste on the one hand, and to political power or traditional kingship on the other' (Dumont, 1966, p. 85). Maine did regard villages as forming part of wider kingdoms but Dumont senses a 'contradiction' in the way in which the relation between the two was formulated: 'The contradiction comes up forcibly in another passage [of Maine]: "[the kings] swept away the produce of the labour of the village-communities and carried off the young men to serve in their wars but did not otherwise meddle with the cultivating societies" '. He concedes that 'Maine and other writers with him are probably right in assuming that kings did not interfere with the principles on which the villages were constituted, and one must distinguish between material or factual interdependence and juridical or moral intervention'. Since 'all over the country the villagers agreed to deliver to the king a substantial part of the produce' they did recognize their dependence on him. And Dumont concludes, 'Such a high degree of factual dependence cannot but be reflected, occasionally at any rate, in the constitution of the village and even in the ideology of its members' (Dumont, 1966, pp. 87–8).

In Section, II, I shall be considering the question of the relation between the king or other higher political entity and the villages, and for the present I merely note that both Marx and Maine lent the weight of their authority to such misconceptions as economic autarky and political autonomy of the Indian village. Since both also thought that communal ownership of land was the pristine practice in Indian villages, it followed from this that villages were equalitarian communities. Though both mentioned caste, they do not seem to have understood how the institution worked or its implications for communal ownership.

The last quarter of the nineteenth century saw an upsurge in nationalist sentiment in India and it is understandable that Indian writers viewed the political and economic changes brought about by British rule from a nationalist angle. The economic exploitation of India, the continuous drain of wealth from India to Britain, the ruin of Indian handicrafts, and the consequent impoverishment of the peasantry became familiar themes with them.[3]

The general result of the nationalist interpretation of Indian economic history has recently been summed up by Dharma Kumar in the following terms:

> British rule, and the flooding of India with foreign manufactures, destroyed domestic industries, and so drove the artisan on to the land. The British introduced in certain areas a system under which land revenue was assessed at high rates and was payable in cash, and which held individuals responsible for payments. This led to the destruction of the old village communities. The British brought about changes in the law which made it possible to sell land; it went either to the State for non-payment of taxes, or to the money-lender for non-payment of debt. This turned the peasant into a landless labourer. Supporters of this theory concede that there were some landless labourers before the British took power, but their numbers, so it is held, were insignificant. (Kumar, 1965, p. 188.)[4]

Gandhi, who placed the peasantry in the forefront of the national consciousness by his political campaigns on their behalf in Champaran (Bihar) and Khaira (Gujarat) districts,[5] felt the need to do something immediately to lessen their poverty and misery. He urged Indians to use hand-spun and hand-woven cloth and converted its

---

[3] See for instance R. C. Dutt, *Economic History of British India under Early British Rule* (London, 1908), pp. 394–420.

[4] However, even during the British Period there were economists such as D. R. Gadgil who took a different view. See his *The Industrial Evolution of India in Recent Times* (O.U.P., London, 1948), pp. 36–42, and 140–1.

[5] See R. I. Crane, 'The Leadership of the Congress Party', in *Leadership and Political Institutions in India* ed. R. L. Park and I. Tinker (Greenwood Press, 1969), pp. 179–80.

wearing into a national cult. It was a part of his *swadeshi* movement and it included the boycott of foreign, especially British, goods. The *swadeshi* movement was a powerful weapon in the hands of a subject people who yearned for their freedom.

Certain elements in Gandhi's world-view made him a strong advocate of the village as against the State and the big city: he was a philosophical anarchist[6] influenced by Ruskin, Tolstoy, and Thoreau. He had a deep suspicion of the power of the State, a hatred of the enslaving machine, and unlike most educated Indians was a staunch believer in the necessity of manual labour for everyone. He was also in the tradition of genuine spirituality, practising and preaching plain living and high thinking.

Gandhi's programme of rural reconstruction involved the revival of handicrafts and panchayats, and the removal of untouchability. He wanted panchayats to arrive at decisions on the basis of consensus, as he was convinced that ordinary democratic processes resulted in the suppression of minority views and interests.

To sum up: The erroneous, idealized, and oversimplified view of the Indian village first propounded by the early British administrators was later cast into the framework of universal history by Marx and Maine. Both of them, Maine more unqualifiedly than Marx, believed in the antecedence of communal ownership over the institution of private property, and Maine actually claimed to have discovered its existence in villages of the North-Western Provinces. It may be noted, in passing, that the theory of the primitive communism of property was popular with nineteenth-century Europeans who saw it everywhere around them. As against this, Marx supported, at least before he came upon Morgan's *Ancient Society*, the theory of the (east) Indian origin of communism in property:

[6] See Geoffrey Ashe, *Gandhi* (New York, 1968). 'He moved toward the view that the State is evil, being coercive in its essence, an organ of privilege. Why should Indians be free from the vices of other rulers? *Better than any existing State would be a copperative federation of village republics'* (pp. 242–3; emphasis mine). See also Gandhi's *Hind Swaraj* (Ahmedabad, 1944): 'I am not aiming at destroying railways or hospitals, though I would certainly welcome their natural destruction. Neither railways nor hospitals are a test of a high and pure civilization. At best they are a necessary evil. Neither adds one inch to the moral stature of a nation. Nor am I aiming at a permanent destruction of law courts, much as I regard it as a "consummation devoutly to be wished". Still less am I trying to destroy all machinery and mills. It requires a higher simplicity and renunciation than the people are today prepared for'. From the introduction ('A Word of Explanation') to the 1921 edition, pp. xi–xii.

A ridiculous presumption has latterly got abroad that common property in its primitive form is specifically a Slavonian, or even exclusively Russian form. It is the primitive form that we can prove to have existed among Romans, Teutons, and Celts, and even to this day we find numerous examples, ruins though they be, in India. A more exhaustive study of the Asiatic, and especially of the Indian forms of common property, would show how from the different forms of primitive common property, different forms of its dissolution have developed. Thus, for instance, the various original types of Roman and Teutonic property are deducible from different form of Indian common property.[7]

The communal ownership of land certainly has equalitarian implications. Marx loathed the Indian village and the kind of life which was possible in it, and he regarded its disappearance as a result of the economic forces and technology introduced by the British, as both necessary and inevitable. It is therefore surprising to find a few Indian Marxists such as A. R. Desai presenting what Dharma Kumar has aptly called 'Golden Age' descriptions of the Indian village (Kumar, 1965, p. 187). But against this there have been many others, Marxists and non-Marxists, who have refused to subscribe to the equality thesis.[8] Indian social anthropologists and sociologists who have carried out field studies of villages since independence have certainly emphasized the existence of caste and other inequalities. In fact, there is a feeling among their colleagues in economics, political science, and history that they have paid too much attention to caste. In view of this it is surprising indeed to find Dumont writing that many modern Indians assume a community to be an equalitarian institution.

I shall consider in the next two sections the concept of the self-sufficiency of the Indian village, first in the political and then in the economic sense. After that I shall take up the question whether the Indian village is a community or just an architectural and demographic entity.

II

Notwithstanding the frequency with which the term 'little republics' was used, neither the British administrators nor Marx or Maine

[7] As quoted in Thorner, 1966, p. 51.

[8] See, for instance, *The Indian Rural Problem*, M. B. Nanavati and J. J. Anjaria (Bombay, 1944), pp. 74–5. See also the monographs on rural life in Gujarat and Maharashtra published in the 1940s by economists and sociologists in the Departments of Economics and Sociology of Bombay University

regarded villages as self-governing in the full political sense. Such criticism as has to be directed against them pertains only to the manner in which they formulated the relationship between villages and the wider, inclusive political system.

It is, however, possible formally to acknowledge the existence of the State while ignoring it in actual discussions of the village community. Thus, while the fact of the payment of taxes by the village to the State finds mention generally, it is stated that but for this payment villages are autonomous.

> To state, as many moderns have done, that 'apart from' this remittance villages were independent and the villagers the owners of the land, will not do. . . . [Elphinstone] went to the crux of the matter when he wrote: 'though under a settled government, it [the village] is entirely subject to the head of the State, yet in many respects it is an organized Commonwealth' (*History*, p. 68). The idealization begins when dependence on the State is forgotten, and the village is considered as a 'republic' in *all* respects. (Dumont, 1966, p. 74.)

It seems unlikely that villagers were entirely indifferent to the fate of the kingdom of which they were a part. They would have had a natural preference for a 'good' king and a distaste for a 'bad' one judged by such criteria as the share of the crop he collected by way of tax, and the effectiveness of the protection he offered them aginst robbers, marauding troops, etc. Apart from this, the fact that occasionally the king or chieftain hailed from a locally dominant caste resulted in his caste fellows in various villages in the kingdom rallying to his aid in a crisis. In medieval Gujarat, for instance, local Rajput chieftains and their allies, the Koli chieftains, fought, for a period of 400 years, the Muslim conquerors who had displaced the Rajput king of Gujarat (Srinivas and Shah, 1960, p. 1377).

The administrators also highlighted the great ability of the village communities to survive temporary disaster, but this again was an exaggeration as Baden-Powell pointed out:

> As to the villages being unchangeable, their constitution and form has shown a progressive tendency to decay, and if it had not been for modern land-revenue systems trying to keep it together, it may well be doubted whether it would have survived at all. No doubt there are cases in which villages have been re-established by the descendants of a former body driven out by disaster . . . but the invitation of the ruler has much to do with the return: he desires to re-establish deserted estates for the sake of his revenue; and old landholders are the best; while an old headman family has

an obvious capacity for inducing cultivators to restore the village. When villages are refounded, it is however just as often by totally different people. (Baden-Powell, i (1892) pp. (171–2.)

While one must conclude that there was a tendency to exaggerate the quantum of autonomy as well as the stability of the villages, the early writers were only trying to characterize a situation in which the lowest level of the political system, viz. the village, enjoyed a considerable measure of autonomy as well as discreteness from the higher levels. It was also far more stable than units at the higher levels. The latest scholar to comment on this phenomenon is Frykenberg. His remarks on Guntur District in Andhra Pradesh during the period 1788–1848 hold good for other regions and periods as well:

> Villages survived forces and innovations of central authority. Institutions above the villages were seemingly much less durable. Struggles for village and district positions took place whenever a new regime sought to enforce its authority; but power at high levels was much more transient and its danger passed away. Perpetual strife, counter-marching armies, and rapid rising and falling of fortunes are said to have occurred as each king tried to spread the umbrella of his authority over the plains. (Frykenberg, 1965, p. 14.)

One aspect of the relationship between the State and the village which writers have generally commented upon is the former's 'extractive' role. According to Marx, the 'structure of government in Asia had consisted from time immemorial in only three departments: that of Finance, or the plunder of the interior; that of War, or the plunder of the exterior; and finally, the department of Public Works [for irrigation]' (Thorner, 1966, p. 40). Maine held a similar view (see p. 46 above).

Maine commented upon further characteristic of the State in the British period: the State, '. . . with few doubtful exceptions, neither legislated nor centralized. The village communities were left to modify themselves separately in their own way' (Thorner, 1951, pp. 72–3).

O' Malley fills in some details on the relation between the pre-British king and the villages:

> Except for the collection of land revenue there was little state control of the villages. The activities of the state did not go further than the primary functions of defence against external enemies, the prevention of internal rebellion, and the maintenance of law and order. The administrative

machinery can scarcely be said to have extended to the villges. . . . The only contact with the villages was by means of local officials having their headquarters in the towns, who were responsible for patrolling of the main routes, the suppression of organized crime, and the realization of the land revenue. So long as it was paid, and so long as there was no disturbance of the peace endangering the general security or outbreaks of crime preventing the safe passage of travellers and merchandise, the villagers were left to manage their own affairs, with headmen and councils of elders to try their petty cases and village watchmen to prevent petty crimes. (O' Malley, 1941, p. 12.)

While Maine and O' Malley were stating an important truth, we shall see later that the relationship between the king and his subjects was more complicated, and that the traditional king performed certain other duties as well. However, this is not to deny that British rule altered fundamentally the relationship between the rulers and the ruled.

The kind of relationship between the State and the village described above had its roots in primitive technology and the related phenonmena of absence of roads and poor communications. In the first half of the nineteenth century, according to Gadgil, 'In most parts of the country roads as such did not exist, and where they did exist their condition was very unsatisfactory' (1948, pp. 3–4). There was an almost complete absence of roads in the Madras Presidency at the beginning of the nineteenth century. According to the Public Works Commissioners appointed by the Madras government,

> . . . nearly the whole of the made roads (so called) are only so far made as to be just practicable for carts. They admit of carts moving in dry weather with light loads at a very slow pace and by very short stages. But by far the greater portion of these roads are unbridged and a heavy shower cuts off the communications wherever the stream crosses a line and they are in many cases so unfit to stand the effects of the wheels while the surface is wet, that in the monsoon months they are out of use except for cattle or foot passengers. (Gadgil, 1948, pp. 3–4.)

Not until British rule was there an attempt to cover the country with a network of roads but their efforts were confined, more or less, to connecting the main towns. Inter-village communication improved, if at all, only incidentally. Only since 1947 has attention begun to be paid to rural roads. But the situation continues to be extremely unsatisfactory. According to J. M. Healey,

> India has the lowest mileage of road per cultivated acre in the world. Large

areas have no access to roads at all. Only 11 per cent of the 645,000 villages are connected by all-weather roads. One out of three villages is more than five miles from a dependable road connection. The isolation of many villages impedes the spread of new attitudes and techniques as well as movements of physical goods. (Healey 1968, p. 168.)

Any portrayal of villages as helpless entities in their relationship with rulers in pre-British India would not be correct. There is here a need to emphasize the distinction between regular or continuing relationships between the village and the State, and individual instances of contact. In the latter situation, any cruelty or injustice could have been practised, whereas in the former there were several constraints on the king's power. I have already mentioned how geography and primitive technology favoured a measure of village autonomy. This was reinforced by the character of the pre-British political system. Any king who wanted to sit on his throne for a period of time had to win, in some measure, the support and goodwill of his subjects living in the villages of his kingdom. Otherwise he was inviting them to be disloyal to him during crises, which were only too frequent. Troops could certainly be sent to make an example of disloyal villagers but such a measure, one suspects, was resorted to only in an extremity. And, apart from other constraints, the king could not always take for granted the loyalty of his troops. Percival Spear writes of villages in the Delhi region during the last days of Mughal rule:

> They [the landowners] acted as the representatives of the body of proprietors, and in the name of the rest of the village. They had first to fix the assessment with the government officials. This in itself required all the qualities of the diplomatist, the statesman and the soldier. If the Government was short of troops and they put on a bold front they might escape payment altogether. By judicious management, such as presents to soldiers mutinous through arrears of pay, they might turn the troops against their commanders, and even receive money for ransoms instead of paying up. If this was not possible they could retire behind their mud walls and defy the officers, hoping that the rains would break, that a party of marauding Sikhs would gallop up or that the troops might be called away before they had time to bring up the artillery. But they must not resist too long and allow the village to be stormed, when all would be lost in the general plunder. (Spear, 1951, p. 123.)

Villages paid some attention to their defence, and in the Delhi region villages of any size surrounded themselves with a mud wall, and had even watch-towers to protect their walls. Neighbouring villages came together to protect themselves against external attack (Spear, 1951, pp. 125–6). Generally, in most parts of India, the dominant peasant castes seem to have provided the pool from which chieftains were recruited, and some chieftains such as Shivaji even graduated to kinghood. An important criterion of dominance was the ability to field a certain number of men for a fight. Violence was an integral element in the tradition of the dominant castes, and the political conditions of pre-British India provided ample opportunities for violence. Further, the successful exercise of violence often resulted in a caste, or a section of it, being able to claim Kshatriya status.

It was the establishment of the *Pax Britannica* that effectively clipped the wings of the leaders of the dominant peasant castes. But even after a century of British rule many a peasant leader continued to have the attitude and outlook of a chieftain of yore. N. Ramarao, during his lifetime a distinguished official of Mysore State, has presented the portrait of such a leader in his enthralling book of memoirs, *Kelavu Nenapugalu* (Ramarao, 1954, pp. 149–50). Even today in Mysore it is not uncommon to refer to a powerful and autocratic village leader as a *pāḷegār*, a term referring to a chieftain of a group of villages.

Collective flight was another sanction available to villagers against oppression. The sanction was rendered more potent by the fact that labour was scarce in pre-British India while land was relatively abundant. There was also the likelihood that the flight of people in one village would have repercussions elsewhere, given the bonds of kindred and caste which frequently cut across villages. According to the Thorners,

> so long as the peasants turned over to the local potentate his customary tribute and rendered him the usual services, their right to till the soil and reap its fruits was taken for granted. Local rulers who repeatedly abused this right were considered oppressive; if they persisted, the peasantry fled to areas where the customs of the land were better respected. As land was still available for settlement and labour was not too cheap, local chiefs had to be careful lest they alienate the villagers. (Thorner and Thorner, 1965, p. 52.)

Villages then, were, not helpless entities but had considerable resources of their own in dealing with higher political powers. An implicit recognition of this is to be found in the prevalence in pre-British India of a form of government which bore a close resemblance to what Lord Lugard termed 'indirect rule'. Tax-farming was an expression of indirect rule and its popularity was due to the fact that it relieved the king and his administration from preoccupation with the day-to-day problems regarding the villages farmed out. That the system had its dangers, grave ones at that, is not gainsaid. The tax-farmers could—and did—mulct the cultivators, and only a fraction of their collections reached the king. Again, the political system of pre-British India offered temptations if not opportunities for tax-farmers to transform themselves into chiefs. But there were also factors inhibiting their rapacity: the milch cows could run away, putting an end to the milk supply. There was also the likelihood of complaints reaching the king about the inhuman exactions. Punishment was likely to be swift and deterrent in such a case, as the act often provided the king with a chance to regain his popularity with the peasantry. Such punishment also conveyed an unequivocal message to other tax-farming officials.

I am aware that the 'joint' villages of the north have been called 'democratic'—in contrast to the 'severalty' villages elsewhere which have been dubbed 'autocratic'—on the ground that a relationship of equality characterized the representatives of the landowning lineages who formed the village council. The severalty villages, on the other hand, were dominated by the hereditary village headman who wielded enormous power. Leaving aside the fact that the democracy of the 'joint' villages did not include the members of the non-dominant castes, the distinction ignores a fundamental similarity underlying both types of villages, viz. the existence of dominant castes in both joint and severalty villages. This meant that the council of the dominant caste, comprising the elders of the different lineages, was important and even a state-appointed headman could not easily ignore its views.

The point I am making is that kings were willing to let villagers govern themselves in day-to-day matters, and wherever a dominant caste existed, its council, on which the leading landowners were represented, exercised power in local matters. The existence of the dominant caste was of greater importance than the fact that tax was collected either through a single hereditary headman or through a body of co-owners. The council of the dominant caste observed certain rules and principles which operated universally, such as respect for the customs of each caste, respect for the principle of hereditary succession including, in certian contexts, primogeniture, respect for males and elders, and for the authority of the head of a household.

In other words, something akin to 'indirect rule' seems to have been built into the political and social structure. And yet, paradoxical as it may seem, the king seems to have performed other functions and duties besides those of collection of taxes and conscription of young men during war. (This fact has not been sufficiently recognized by the earlier writers.) A good king paid attention to the condition of his people—he built roads, tanks, ponds, and temples, gave gifts of land to pious and learned Brahmins.[9] Disputes regarding mutual caste rank were ultimately settled by him. Such a function was not restricted to Hindu kings: under the Mughal emperors, the Delhi court was the head of all caste panchayats, and questions affecting a caste over a wide area could not be settled except at Delhi (Srinivas, 1966, p. 41).

A good king also paid attention to the development of irrigation, though this is more evident in some parts of the country than in others. In parts of modern Mysore, Tamil Nadu, and Andhra Pradesh both canals and tank-systems have played a significant part in agriculture. According to K. A. Nilakanta Sastry, doyen among South Indian historians,

> the importance of irrigation was well understood from early times; dams were erected across streams and channels taken off from them wherever possible. Large tanks were made to serve areas where there were no natural streams, and the proper maintenance of tanks was regularly provided for. The extension of agriculture was encouraged at all times by granting special facilities and tax concessions for specified periods to people who reclaimed land and brought it under cultivation for the first time. (Nilakantha Sastry, 1966, p. 328.)

A feature of agriculture in many parts of South India is the damming up of rain water in suitable places to form artificial lakes which were constructed in such a way that the overflow water from each tank fed one below till the excess eventually reached a stream. Many of these tanks were large indeed, providing irrigation water for hundreds of acres of land lying on the other side of the embankment. The maintenance of these tanks was an important duty of the king's. Irrigation tanks were also found in other parts of the country such as Gujarat, Malwa, and Bundelkhand, though nowhere were they as numerous (O' Malley, 1941, pp. 233–41).

Thorner has, however, argued that ' . . . canal networks have never

[9] See in this connection T. V. Mahalingam, *South Indian Polity* (University of Madras, 1955), pp. 26–31; see also R. C. Dutt, 1908, pp. 196 ff.

been the outstanding feature of Indian crop production. Rather, Indian agriculture as a whole has always turned on rainfall and the local wells or ponds of the villages' (Thorner, 1966, p. 40). If Thorner had looked below the all-India level to the regional, he would have found that irrigation, through canal and tank, was an important feature of pre-British agriculture in certain parts of the country. The bigger irrigation projects could not have come into existence without the king's active support and involvement.

The gist of my argument is that the relationship between the king and the village in pre-British India was a complex one. The villages were not without some resources in any continuing relationship between them and the king. The king's functions were not confined to collecting taxes and conscripting young men during war. He had also other duties.

Dumont has argued that Maine and other writers failed to understand the implications of the regular collection by the king of a 'substantial part of the produce'. This meant, in effect, that 'wherever the king delegated his right to one person there was a chance of this beneficiary and his heirs assuming the superior right and reducing its former enjoyers to subordinate status' (Dumont, 1966, p. 88). In other words, the king's power was effective enough to ensure that the rights of individuals who were recognized by him prevailed over other rights.

The payment of a substantial portion of the produce was then also a symbol of the village's dependence on the king. Dumont quotes Maine himself to make the point that Indian villagers exhibit their 'dependency' on the State by the importance they attribute to 'the sanction of the state, be it only in the form of the stamped paper on which an agreement between private parties is written' (Dumont, 1966, p. 88). My own field experience supports this. In 1948, I was surprised to find Rampura villagers frequently mentioning the existence of copper-plate grants listing the privileges, duties, and rank of particular castes. The copper plates were 'somewhere', with someone not available, near by; but what struck me was the fact of their being mentioned.

III

The myth of economic self-sufficiency or autarky of the pre-British village is one that is widely subscribed to, and it has persisted until very recent years. But no one, not even Dumont, has drawn attention to the contradiction between the fact of the subsumption of the village in the wider polity and the notion of its economic self-sufficiency. The former substracted from self-sufficiency in that the State continually drained away a sizeable share of what was produced and left the village to make do with what was left over. There were also others such as tax-farmers and officials who came in for their share. And this was during times of peace. War not infrequently destroyed, at least temporarily, the economy of the village.[10]

However, it is not surprising that observers of the Indian village have been impressed with its appearance of economic 'self-suffi-ciency'. The crops provide food and seeds for the next season, taxes to the State, and the means to pay essential artisan and servicing castes such as the carpenter, blacksmith, potter, barber, washerman, and priest. In an economy which is non-monetized or minimally monetized, where poor communications confine the flow of goods and services to a limited area, the wants are few and are such as can be satisfied locally. The appearance of self-sufficiency was enhanced by caste-wise division of labour.

A closer look at the village will, however, reveal several loopholes in self-sufficiency. Even a basic commodity like salt was not produced in most villages, and many spices also came from outside. Iron, indispensable for ploughs and other agricultural implements, was not available everywhere, and iron-smelting was a localized industry. Sugar-cane was not grown in all villages (Gadgil, 1948, p. 60),[11] and it was the biggest source of jaggery, widely used by the peasantry. Betel leaves and areca nuts, coconuts, tobacco, and lime paste were other peasant wants not always locally met. Silver and gold were essential for wedding jewellery, and they had to be imported from the towns. And not every village had goldsmiths.

Weekly markets are a feature of rural India everywhere and they are a traditional institution. They dramatize the economic inter-

[10] Buchanan, for instance, mentions that a large tract of the country to the north of Tippu's capital of Seringapatam in Mysore had been laid waste at the time of Lord Cornwallis's invasion in 1792: '. . . the people had been forced by Tippu Sultan to leave the open country and retire to the forests where they lived in huts and procured provisions as best they could. A large proportion of them had perished of hunger, and the country was only sparsely populated even in 1800 . . . .' (Dutt, 1908, p. 210).

[11] After making the usual bow to economic self-sufficiency, Gadgil proceeds to observe, 'There were only two important kinds of agricultural produce which, on account of their nature, could not be grown generally all over India . . . . These were cotton and sugarcane. The trade even in these was of a limited extent and the area it covered was also limited.'

dependence of villages and provide conclusive refutation of the idea of economic self-sufficiency.[12] It is indeed surprising that their existence has been ignored by most writers. The areas serviced by weekly markets seem to have varied from market to market, many having more than a purely local reputation. There seems to have been also a degree of specialization in weekly markets on the basis of the goods sold there.

The periodical fairs held on the occasion of the festival of the local deities or on certain sacred days (e.g. the full moon in Kārtik or Chaitra) were also visited by villagers in large numbers, and the fairs served many purposes, secular as well as religious. In southern Mysore, for instance, the annual fairs held in Chunchanakatte, Hassan, Mudukutore, and Madeshwara hills were well known for the buying and selling of cattle.

The concept of economic self-sufficiency also assumes that every village had living within it all the essential artisan and servicing castes. There can be some argument as to which are the essential castes, but those who have first-hand knowledge of rural India would probably agree that peasants would have a continuing need for the services of the carpenter, blacksmith, leatherworker, potter, barber, and washerman. This would mean that every village had to have at least seven castes. This was—and still continues to be—highly unlikely. The number of castes in a village is related to its total population, and according to Karan, 'In the north they [villages] are small with an average population of about 500; in the south they are large with nearly 1000 inhabitants. About one-fourth of all Indian villages have less than 500 inhabitants, another quarter have populations exceeding 2000 and the rest fall in between' (Karan, 1957, p. 56).

Kingsley Davis quotes from a survey, carried out in the early 1930s, by S. S. Nehru, of fifty-four villages in the middle Ganges valley, and Nehru found fifty-two castes inhabiting the area.

> Not one of these castes, however, was represented in every village. The Chamars, one of the most pervasive, were found in only 32, and the Ahirs in only 30 villages. 'And yet, *a priori*, the Chamars should be represented in all villages, as they are the commonest type of *razil* population, supply all the

[12] Rivett-Carnac's description of a weekly market in Chinmoor in the former Central Provinces has been quoted by Gadgil, 1948, pp. 308–9. The date of the description is 1867–8, 'just after road building had vigourously begun'. Granting that a market such as this would have included fewer articles and attracted fewer persons before vigourous road-building, still its existence provides eloquent evidence of inter-village economic interdependence.

labour in the village and are indispensable to village life'. There were Brahmins in 40 per cent of the villages. The Nai, or barber caste, was represented in less than half the villages. The average number of villages in which each caste, taken altogether, was represented was only 9·3.

Davis concludes that

> ... the rural village has by no means a full complement of castes, and that the castes it does have are generally represented by one or two families. The first fact means that each village must depend to some extent upon the services of persons in other villages, and the second that relations between caste members must be maintained by contact between villages. (Davis, 1951, p. 166.)

It is likely that the proportion of smaller villages was greater in pre-British India, for it was during British rule that large irrigation projects were undertaken in different parts of the country. And irrigation enables larger numbers of people to be supported on the same quantum of land through the intensive cultivation of more profitable crops. Irrigation increases the demand for labour and puts more money into the pockets of peasants; the money then becomes translated into new wants to be satisfied by new goods and services.

Individual villages, it is clear, are far from self-sufficient economically. It may be added that socially and religiously, also, villages were anything but self-sufficient. Caste ties stretched across villages, and in a great part of northern India the concept of village exogamy, and the existence of hypergamy on a village basis, constitute an advertisement for inter-village interdependence. The partiality of peasants for pilgrimages and fairs also highlights the fact that the Indian village was always a part of a wider network.

IV

Dumont and Pocock have argued that the Indian village is only an 'architectural and demographic fact' and that fieldworkers confer upon the 'village a kind of sociological reality which it does not possess'. Further, 'the substantial reality of the village deceives us into doing what we normally would not do in a social analysis and into assuming *a priori* that when people refer to an object by name they mean by that designation what we ourselves mean when we speak of it'. Dumont and Pocock argue that the 'village' has a different meaning for rural Indians: 'Whether a man is speaking of his own village or of another village, unless he positively specifies another caste by name, he is referring to his caste fellows' (1957, p. 26). 'Village solidarity', which many anthropologists, including myself, have reported as a reality is nothing else but a 'presupposition imposed upon the facts' (1957, p. 27). The lower castes do not possess a sense of loyalty to the village. They are clients of powerful patrons from the dominant caste(s), and the obligations of clientship force them to act in ways which are misinterpreted as arising out of 'village solidarity'. Speaking of Rampura in particular, Dumont and Pocock have asserted that 'if the solidarity of the village means anything it is the solidarity of the local group of Okkaliga [dominant caste]' (1957, p. 29). Caste, and occasionally, factional divisions, are so fundamental that they make a local community impossible.

As already mentioned, Dumont is critical of Indian scholars for ignoring the inequality inherent in Indian villages:

> Caste is ignored or underplayed throughout, for in the prevalent ideology of the period a 'community' is an equalitarian group. This characteristic gains in importance as the conception spreads, becoming more and more popular. Dominance, and even hierarchy, are not absolutely ignored by all writers, but they do remain on the whole in the background, and the main current of thought sustained by the expression 'village community' goes against their full recognition. Indeed, the question arises whether this is not finally the main function of the expression. (Dumont, 1966, p. 67.)

If Indians have used the term 'community' to ignore inequality, to Dumont Indian villages are not communities because of the inequality of caste. He does not consider at all the question whether unequal groups living in small face-to-face communities can have common interests binding them together. It is assumed implicitly that equalitarianism is indispensable to community formation, and also that such communities are the rule in the Western world, or at least in Western Europe. There is no reference to the existence of economic and social inequalities in Western Europe. The assumption seems to be that when inequalities assume the form of caste they make community impossible.

In order to understand the part played by caste in the local community it is necessary to place caste against the background of the demographic, political, economic, and ideological framework of pre-British India. I shall now briefly list a few significant features of caste in pre-British India. In the first place, caste was generally accepted as well as ubiquitous. The idea of hierarchy, for the division of society into higher and lower hereditary groups was regarded as natural, and caste-like groups were to be found even in sects which rebelled against Hinduism, and among Muslims and Christians.

The demographic situation in pre-British India affected inter-caste relations in significant ways. Kingsley Davis has postulated that India's population remained more or less stationary 'during the two thousand years that intervened between the ancient and the modern period', and that 'the long-run trend would be one of virtual fixity of numbers. No real change could have occurred in this condition until the coming of European control, and then only slowly' (Davis, 1951, p. 24). One result of this demographic stagnation was the relative ease with which land was available for cultivation. According to Spear, reclaimable waste land was 'plentiful in the eighteenth century' in the Delhi region (Spear 1951, p. 118). Extensive areas were available for cultivation in Madras Presidency at the beginning of the nineteenth century, and this was due to several factors such as the political instability of the preceding decades, high land revenue, and the existence of virgin lands (Kumar, 1965, pp. 107–8).

Maine has commented on the effects of a historical situation in which land was plentiful relative to available labour:

> Right down to the last few generations there persisted a singular scarcity of indigenous law pertaining to tenures. Men remained of more value than land, the village communities continued to absorb outsiders in the hope of getting more land tilled and meeting the burden of revenue payments to political potentates, monarchs or emperors. The need for additional culti-vators helped preserve a power of elasticity and absorption in the villages and kept them from becoming with any rapidity, closed corporations. (Thorner, 1951, pp. 73–4.)[13]

The net result was a situation where landowners competed for the services of labourers, the exact opposite of that prevailing today in large and irrigated villages. The highly institutionalized nature of the employer–labourer relationship in pre-British India may well have represented an effort on the part of landowners to assure themselves of a steady source of labour. Such institutionalized relationships were characteristic of the country as a whole and not merely of a part. According to Daniel and Alice Thorner,

> like other elements of the Indian agrarian structure, the relation between employers and labourers takes numerous diverse forms. We find in the literature hundreds of indigenous terms, each denoting a particular kind of

[13] As summarized by Thorner.

labour or labourers. Even for a single such word, the terms of employment, duration of the work, amount and form of payment may vary from district to district and village to village. (Thorner and Thorner, 1965, p. 21.)

The existence of institutionalized master–servant relationships not only assured each landowner of a steady source of labour but also helped to minimize competition between landowners for labour which might otherwise have split them into rival factions. There were also other factors which kept down factionalism. The threat to property and life from rapacious chiefs, freebooters, and dacoits and from such natural calamities as flood, famine, and epidemics emphasized the common interest of all villagers in sheer survival. It follows from this that British rule, bringing in law and order, and welfare measures, favoured the development of factionalism and at the expense of the village.

Strong employer–employee bonds provided a countervailing force to caste since generally the employers came from a high or dominant caste while the landless labourers generally came from the lowest strata. Dharma Kumar has observed that

one of the most striking and important peculiarities of the Indian forms of servitude is their close connection with the caste system. Most types of servile status were hereditary, and in general the serfs and slaves belonged to the lowest castes. Although this group as a whole was at the bottom of the caste ladder, there were further gradations within it, each sub-group having its articulated rights and liabilities. In fact, the caste system not only confirmed the economic and social disadvantages of the agricultural laborer, but also gave him some rights, some economic, others of a social and ritual nature. (Kumar, 1965, p. 34.)

There is no reason, however, to think that landless labourers were always confined to the lowest castes, and it is likely that one of the results of the great population increase of the last hundred years or more has been the reduction of many individuals from the landowning hierarachy to the lower levels of peasantry. Here, the customary patterns of expenditure prescribed for wedding and funerals, and the hazards to which the subsistence agriculture of poor cultivators are subject, as, for instance, failure of the crops following drought or the death of a plough bullock during the agricultural season, played their part in pushing marginal landowners down the economic slope. In any case, in rural India today landless labourers hail from a variety of castes, high and low, though it is even now true that the lowest castes generally provide the bulk of such labourers. Quite frequently, poor members from the dominant peasant castes are found serving their richer caste fellows as labourers and servants. It is only the Brahmin who does not substantially contribute to the ranks of landless labourers, this being particularly true of peninsular India.

In short, both political and economic forces in pre-British India converged to put a premium on localism, and to discourage the formation of horizontal bonds stretching across political boundaries.[14] Movement across political boundaries was difficult for all, and especially for members of the lower castes. Besides, politically ambitious patrons had to keep in view the paramount need to acquire and retain their local followers. They had to be generous with food and drink, especially at weddings and funerals, and provide loans and other help when necessary. Also the fact that it was difficult to store grain or other foodstuffs for long periods in a tropical climate made necessary the distribution of surpluses. The far-sighted leader who gave, or loaned on interest, foodgrains to his tenants and labourers earned their goodwill, which could be cashed in on a later occasion. Thus political factors combined with the ecological to favour an ethic of distribution which in turn was buttressed by ideas from the great tradition.

The tendency to stress intra-caste solidarity and to forget inter-caste complementarity is to ignore the social framework of agricultural production in pre-British India. Castewise division of labour forced different castes living in a local area to come together in the work of growing and harvesting a crop. Landowners forged inter-caste ties not only with artisan and servicing castes but also with castes providing agricultural labour. These last-mentioned ties involved daily and close contact between masters from the powerful dominant castes and servants from the Untouchable or other castes just above the pollution line. Again, in the context of a non-monetized or minimally monetized economy, and very little spatial mobility, relationships between households tended to be enduring. Enduringness itself was a value, and hereditary rights and duties acquired ethical overtones.

I have subsumed institutionalized relationships between a land-

[14] See in this connection E. Miller's 'Caste and Territory in Malabar', *American Anthropologist*, 56 (1954), 416–17: 'Movements of the military Nayar subcastes were similarly circumscribed by political boundaries. For all lower castes the chiefdom was the limit of social relations within the caste, while their relations with other castes were largely confined to the village' (p. 416).

owner and his labourers, and between him and households of artisan and servicing castes, in a single category, viz. patron and client. A characteristic of this relationship was that it became multi-stranded even if it began as a single-stranded one. It was such a strong bond that it attracted to itself a code of norms, and occasionally, the exceptionally loyal client was buried close to his patron's grave.

Thus, given a situation in which labour was scarce in relation to the available land, village society was divided into a series of production pyramids with the landlord at the apex, the artisan and servicing castes in the middle, and the landless labourers at the bottom. Rivalry between patrons was minimized by institutionalized relationships, and by the existence of external threats to the village community as a whole. A politically ambitious patron had to break through institutionalized arrangements if he wanted to achieve power at the higher levels of the political system, which was characterized by fluidity.

In pre-British India, both technological and political factors imposed limitations on the horizontal stretch of castes, while castewise division of labour favoured the co-operation of households from different castes. The relative scarcity of labour and the institutionalization of the master–servant relationship resulted in forging enduring bonds, between households of landowners and landless labourers, hailing from different castes. Spear accurately characterizes the situation when he refers to ' . . . the classes which, locked by economic, social and religious ties into an intimate interdependence, made up the village community' (Spear, 1951, p. 123). A most significant effect of British rule on the caste system was the increase in the horizontal solidarity of individual castes and the facilitation of their release from the local multi-caste matrix.

V

The members of a dominant caste are in a privileged position *vis-à-vis* the other local castes, and its leaders wield considerable power. These leaders have the greatest stake in the village, have command over resources and, generally, it is they who organize local activity, whether it be a festival, general protest, or fight. They dominate the traditional village council or panchayat.

However, the power exercised by the leaders of the dominant caste traditionally has been subject to some of the constraints to which the king's was in pre-British times. The leaders were required to show respect for certain values common to all castes, and for the customs of each caste even when they differed significantly from those of the dominant castes.[15] In addition, each leader of the dominant caste was bound by strong ties to his clients, and it may be assumed that he was not entirely impartial when matters affecting them came up before the council. The village council, then, had to acknowledge the existence of certain rules and principles, and also, of certain checks and balances which came into play when cases concerning non-dominant castes were being arbitrated upon. There was, however, considerable room for manoeuvre, and this not unnaturally provided ground for charges of corruption, favouritism, etc., against the members.

The leaders of the dominant caste were expected to protect the interests of the village as a whole and were criticized if they did not. I have described elsewhere an incident which occurred in March 1948 in which the headman of Rampura played a leading part in preventing the government from taking away the villagers' right to fish in their tank (Srinivas, 1955a, p. 25). A petition was despatched to the government stating that villagers had fished without hindrance from time immemorial, and that the government's sudden decision to set aside this right by an order was arbitrary. When the government ignored the petition and fixed a date for auctioning fishing rights, word was passed round to everyone, including leaders from neighbouring villagers, not to offer bids at the auction. The boycott was successful, and the villagers experienced a sense of triumph.

In the above instance, the lead was certainly taken by the headman and other members of the dominant caste, but the matter affected the entire village, and in particular, the non-vegetarian castes. The idea is implicit that the leaders of the dominant caste have to work for the village as a whole and not for advancing their personal interests. It may be that the idea is more often respected in the breach than in the observance but that is a different issue.

Dumont and Pocock have also argued that 'even when disputes occur between "villages" it would appear from the evidence that these could be more appropriately described as disputes within the Okkaliga caste about land upon which the Okkaliga base their superiority' (1957, p. 29). In the first place, not all disputes between villages refer to land, and this is true of other villages besides Rampura. For instance, a big fight, in which armed police had to be called, occurred between Kere and Bihalli in October 1947 during the

[15] See in this connection my 'The Dominant Caste in Rampura', *American Anthropologist*, 61, 1 (1959), 1–16.

annual festival of the deity Madeshwara, and the subject of the dispute was the right to carry the portable icon of the deity in procession round the temple. There was a long queue and the party of Kere youths carrying the icon were asked to hand it over to others after their second trip round the temple. The youths replied that they would do so only after completing the third trip. There is a belief that odd numbers are auspicious while even ones are not. A few Bihalli youths tried to wrest the icon from the Kere Youths, and a fight ensued involving injuries to several and premature closure of the festival. The dispute was settled only six months later, and Rampura and Hogur leaders played an important role in the tortuous peace negotiations. While it was true that Okkaligas took the lead in the fight, the fighters were convinced that they were fighting for the honour of their villages.

Even more interesting is the fact that Kathleen Gough has reported the occurrence, over a period of twenty years, of four instances of fights between the lower castes in Kumbapettai and neighbouring villages. Members of the dominant caste of Brahmins were not involved in any of these fights (Gough, 1955, p. 46). It is surprising that this observation should have escaped Dumont and Pocock. Similarly, referring to the highly factionalized village near Delhi studied by Oscar Lewis, they raise the question, ' . . . it would be interesting to see to what extent Oscar Lewis' account of the lack of even this kind of solidarity [of the local dominant caste group] is applicable elsewhere' (Dumont and Pocock, 1957, p. 29). But according to Lewis, even extreme factionalism did not prevent his field-village from acting as a unit on occasion: 'As we have seen earlier, there are occasions when the village acts as a unit. However, these relatively infrequent and with the weakening of the old and traditional *jajmani* system the segmentation within the village is all the more striking, nor has it been replaced by any new uniting forms of social organization' (Lewis, 1951, p. 32). The ability of a village comprising a large number of castes, and also divided into factions, to function on certain occasions as a unit has impressed many observers. A. C. Mayer, for instance, is able to state, 'This account has shown how it is that a village containing twenty-seven different caste groups, each with its barrier of endogamy and often occupational and commensal restrictions, can nevertheless exist to some extent as a unit' (Mayer, 1960, p. 146).

It is possible for villages to function as untis in spite of the various cleavages within them because everyone, irrespective of his caste and other affiliations, has a sense of belonging to a local community which has certain common interests overriding caste, kin, and factional alignments. It is likely that loyalty to the village was greater in the past than now, and future developments may weaken it even further. But the important fact is that it does exist in some measure today. Indian villagers have a complex system of loyalties: in an inter-caste context, identification tenc's to follow caste lines and this is often reinforced by castewise division of labour. In an intra-castes situation, on the other hand, affiliation ollows village lines. This is dramatized in Rampura, for instance at weddings in the ritual of the distribution of betel leaves and areca nuts (*doḍḍa veelya*) to the assembled guests. Besides each guest receiving a set of betel leaves and areca nuts in his role as kinsman or casteman, some receive it also as representatives of their respective villages. The credentials of a person to represent his village are not always clear, and a man whose credentials are rejected feels humiliated. A few guests may claim precedence for their villages over the others, and this has often led to a heated debate. In 1948, elderly villagers spoke feelingly about the difficulties and dangers with which ritual betel-distribution was beset. The elders of Rampura had previously passed a rule which enabled anyone who paid Rs 8.25 to the village fund to escape the cost and trouble of the formal distribution of betel.[16]

In the foregoing pages I have tried to adduce evidence to show that the village is not only an architectural and demographic entity but also a social entity in that all villagers have some loyalty to it. Rural Indians live in a system of complex loyalties, each loyalty surfacing in a particular context. It is presumable, indeed likely, that there are occasions when there is a conflict of loyalties but that is a different matter. The phenomenon of the dominant caste is indeed impressive, but it is not the whole story.

VI

The exclusion of Harijan castes from access to wells and temples used by the others is occasionally cited in support of the argument that the village does not really include all those living in it. It is relevant to point out in this connection that in traditional India relationships between people took place in an ideological framework that accepted caste. A person was born into a caste which was a unit in a hierarchy to castes, and relations among these were governed by the ideas of

[16] See in this connection Srinivas (1955a) pp. 32–3.

pollution and purity. This meant that in actual social life, the highly elaborated and systematized principles of inclusion and exclusion of individuals on the basis of caste came into play. Individuals from a particular caste were included in one context (e.g. living together) and excluded in another (e.g. endogamy), and this applied all along the line. Inclusion and exclusion were also matters of degree—for instance, the social distance between castes in Kerala was traditionally expressed in spatial terms.

The position occupied by a caste in the local hierarchy is not always clear, and a caste's own conception of its position frequently differs from that assigned to it by its structural neighbours.[17] In Rampura, for instance, each caste, including the Harijan, is able to point to another as its inferior. Thus the Harijan Holeyas are able to point to the Smiths as their inferiors, and as 'evidence' of their superiority, to the fact that they do not accept food and drinking-water from Smiths. They would also point to the Smiths being included in the 'Left-hand' castes while they themselves were included among the 'Right-hand' castes. Again, traditionally Smiths were not allowed to perform weddings within the village and were subjected to certain other disabilities. On their side, the Smiths would dismiss the Harijan claim as absurd, and point to their Sanskritized style of life, and to their not accepting food cooked by any except Brahmins and Lingayats as evidence of their high status. This kind of ambiguity regarding the position of several castes was not only a function of the flexibility of the system but also facilitated its acceptance.

I have stated that exclusion was not only contextual but also a matter of degree. An interesting instance of this has bearing on the question of membership of the village. One of the most important temples in Rampura—certainly, architecturally the most striking—is dedicated to the Sanskritic deity, Rāma. While its priest is a Brahmin, the temple is maintained by contributions from the entire village, and the headman's lineage not only has made substantial contributions in the past but also takes an active interest in the temple's activities. It came, therefore, as a surprise to me to find the headman (along with other members of the dominant Okkaligas) refusing to enter the *sanctum sanctorum* (in Kannada, *garbha guḍi*) of the temple while he urged Brahmin devotees to stand there while the *puja* was being performed. Had he wanted to enter the *sanctum sanctorum* no one,

[17] See in this connection Srinivas (1962).

certainly not the Brahmin priest, who was heavily dependent upon him, would have prevented him. But the headman chose to remain outside. It did not occur to him that his full participation depended on his being on a par with Brahmin devotees. Given this kind of ideology, it should be clear that the exclusion of the Harijans from the temple cannot be interpreted as meaning exclusion from the village. The same can be said for other instances of exclusion of Harijans. Condemnation of the exclusion of Harijans (or any other caste) from the point of view of a new ethical system is a distinct phenomenon, and as such ought not to be brought in while trying to understand the meaning of exclusion and inclusion in the traditional system.

Again, exclusion has different connotations in different situations. For instance, in Rampura, four groups of villagers, viz. Brahmins Lingayats, Harijans, and Muslims, are not called upon to make any contribution to the expenses of the Rāma Navami, the great annual nine-day festival at the temple of Rāma. Not only are the Brahmin, Lingayat, and Smith[18] households exempted from making contributions: they also receive the raw ingredients of a meal as their caste rules prevent them from eating food cooked by Peasants. The Lingayat may be regarded as a kind of Brahmin in view of his staunch vegetarianism and teetotalism, and in view of the fact that he acts as priest in two important village temples. In Sanskritic Hinduism, a gift made to a Brahmin confers religious merit on the giver. Not accepting contributions from the Brahmin, and giving him the ingredients of meal, may be interpreted as conferring merit on those who make the contributions. No such consideration is applicable in the case of the Harijan or Muslim.

Harijans, however, perform many essential services at the festival. The hereditary servants run errands for the organizers, and Harijan women clean the rice and lentils for the ninth-day dinner. Harijan men, beating tom-toms, march at the head of the deity's procession. They also remove the dining-leaves on which the villagers eat the dinner. They are the last to eat.

Muslims also participate in the procession. In Rampura a fireworks man, invariably a Muslim, follows Harijans, setting off fireworks. In 1948 one Muslim youth distinguished himself at the procession by a brilliant display of swordsmanship. The fact is that everyone looks forward to the procession with its music, fireworks,

[18] It appears as though the Smiths have won recognition in Rampura as a high caste. I do not know if this is also true in other villages nearby.

display of sword and stick (*lāthi*), fancy dress, and the monkey-god who walks on rooftops peeling green coconuts with his teeth and hands.

The traditional tasks performed by Harijans at village festivals had begun to be regarded by the middle of the twentieth century as degrading symbols of untouchability, and some Harijans attempted to refuse performing them. But the dominant castes used the twin sanctions of economic boycott and physical force to coerce them into conformity.

It is obvious that exclusion and inclusion need to be viewed over the whole range of contexts, religious as well as secular, in understanding the position of a caste. Thus, groups excluded in religious contexts may have important roles in secular contexts. For instance, individual households from the dominant or other high castes often have storng economic ties with households of Harijans. Thus, traditionally, each Harijan household served as 'traditional servant' (*halé maga*) in a peasant or other high-caste household.

> This traditional servant had certain well-defined duties and rights in relation to the master and his family. For instance, when a wedding occurred in the master's family, then men of the servant family were required to repair and whitewash the wedding house, put up the marriage *pandal* before it, chop wood to be used as fuel for cooking the wedding feasts, and do odd jobs. The servant was also required to present a pair of leather sandals (*chammāligé*) to the bridegroom. Women of the servant's family were required to clean the grain, grind it to flour on the rotary quern, grind chilies and turmeric, and do several other jobs. In return, the master made presents of money and of cooked food to the servant family. When an ox or buffalo died in the master's family, the servant took it home, skinned it, and ate the meat. He was required, however, to make out of the hide a pair of sandals and a length of plaited rope for presentation to the master. (Srinivas, 1955b, p. 27).

The institution of *jita* or contractual servantship was an important institution in which the servant worked long hours on the master's farm for a stipulated annual fee (Srinivas, 1955b, pp. 27–8). While many Harijans worked as *jita* servants for high-caste, landowning masters, all *jita* servants were not Harijans nor were all Harijans *jita* servants. Traditionally, *jita* servants did a great deal of the hard work on the land and they were indispensable to the economy. The situation described by Spear for the Delhi region in the early nineteenth century highlights the importance of the castes which normally provided the labour on the farm in contrast with the Brahmin priest whose services were symbolic:

> Thus the lowly Chamar, the cobbler and dresser of unclean leather, received the highest allowance of all, while the priest or Brahmin was given by these hard-headed people the least. He had to make up as best he could by exacting presents on occasions like marriages and deaths when his presence was essential, and by soliciting gifts at festival times, when people were good-tempered and liked to stand well with the gods. ... The barbers and the water-carriers, two other despised occupations, were also rated highly. (Spear, 1951, p. 121

To sum up: In a society where the ideology of caste is fully accepted, and where the principles of exclusion and inclusion apply to everyone inluding the highest, the exclusion of a caste from particular contexts cannot be adduced as evidence of non-membership of the local community. To do so would be to misinterpret indigenous behaviour.

VII

I have so far looked at village unity or solidarity from the outside, and shall now turn to the villagers' perception of the problem. As it happened, I stumbled on to this while I was carrying out a census of village households in Rampura. From the villagers' point of view two factors seem to be crucial in determining who belonged and who did not: length of stay in the village, and ownership of real property, especially land. If a family had spent two generations in the village, and owned a little land and a house, then their membership had been establised beyond question. Surprisingly enough, membership did not seem to have any connection with caste or religion. Exclusion from various activities was not viewed as being relevant to membership : it was something incidental to living in a caste society.

The question of membership in the village came up in the case of three groups, Basket-makers, Swineherds, and Muslims. There were seven basket-makers in Rampura in 1948, and all of them, men as well as women, lived by making artefacts with bamboo bought in the market in Mysore. They made essential articles such as baskets, fish traps, partition screens, and winnows, and sold many of their products in the weekly markets around Rampura. Unlike the other village artisans they worked strictly for cash. They had the habit of never delivering goods on time, and this and their fondness for liquor had given them a reputation for undependability.

More importantly, all the Basket-makers were immigrants from the nearby town of Malavalli, and every few years a family would pack up and return home and its place would be taken by another. Only one household had spent about twenty years in Rampura. None of the Basket-makers owned any real property—indeed, all of them lived on the verandas of houses owned by others. Veranda-dwelling (*jagali mēlē wāsa*) symbolized extreme poverty, and it was generally accompanied by the dwellers' doing casual labour (*kooli kamblạ*) for a living.[19]

While the Basket-makers made articles essential for the villagers, they were not regarded as properly belonging to the village. The Swineherds were another marginal group and had settled down in Rampura during the headman's father's days. They spoke among themselves a dialect of Telugu which was unintelligible to the others. During the rainy season they lived in a cluster of huts on the headman's mango orchard, and during the dry season moved to temporary huts erected on the headman's rice land below the C.D.S. canal. Their transhumance was a tribute to the power wielded by the headman's father: it was his way of ensuring that two of his fields were fertilized by pig manure.

The Swineherds had a herd of about sixty pigs, and a boy took them out 'grazing' every day. The pigs were scavengers and the low ritual rank of the Swineherds was due to their association with pigs. They also ate pork and drank toddy. Their women went round the villages begging and telling fortunes.

The Swineherds had regular relations with only the headman's household. They were tenants of the headman, cultivating 2·5 acres of his land. They occasionally borrowed money from him, and sometimes took their disputes to him. The rest of the village did not interact with them except when one of their pigs damaged plants in someone's backyard or field. They seemed to be in the village but not of it.

As a group, neither Swineherds nor Basket-makers were integrated with the village in the way the Harijans were. But Basket-makers, Swineherds, and Smiths all made contributions to the Rāma Navami festival, while Harijans, along with Brahmins, Lingayats, and Muslims, did not.

The Muslims were the third biggest group (179) in Rampura in 1948, and though they were scattered all over the village there was

[19] In 1948, out of a total population of 1,523, 31 lived on verandas, 14 being widows, 1 a divorcée, and the rest remnants of once fuller households.

some tendency for them to cluster along the fringes of the high-caste areas. They were engaged in a variety of occupations, agriculture, trade—especially trading in ripe mangoes—and in crafts such as tailoring, tinkering, shoeing bullocks, and plastering. Many of them had migrated into Rampura in the 1940s but there was also a nucleus of earlier immigrants who owned land and houses, and did a little buying and selling of rice on the side. It was when I was carrying out a census among Muslims that I heard villagers use two expressions which I came to realize were significant: the recent immigrants were almost contemptuously described as *nenné monné bandavaru* ('came yesterday or the day before') while the old immigrants were described as *ārsheyinda bandavaru* ('came long ago') or *khadeem kulagalu* ('old lineages'). Only three households fell into the latter category and all of them owned land.

It was at this point that I realized the crucial importance of ownership of arable land in determining membership in the village. Villagers are acutely aware of the many ways in which land provides bonds with the village, and land once acquired cannot be disposed of easily, for public opinion is against the disposing of such an important and respected asset. Ownership of land enables owners to have enduring relationships with others in the village: they are able to pay the artisan and servicing castes and labourers annually in grain. Such payments are made on a continuing basis. The owners achieve the coveted status of patrons.

In their talks with me, my friends occasionally referred to such-and-such a Brahmin family as belonging to Rampura even though it had left the village two or three decades ago to settle down in Mysore or Bangalore. But the important fact was that they continued to own land in the village, and collected their share of the crop regularly. This kept alive their links with the village. I may add here that during the 1950s I talked to several Brahmin landowners in Mysore and Bangalore and all of them seemed to feel that land was not only economically profitable but also a link with the ancestral village. Selling land was not only improvident but almost implied a lack of piety toward the ancestors who had acquired the land with great hardship.

During the summer of 1952, when I visited the village a second time, a Shepherd sold all his land to settle down in his wife's village. While conflicts among grown brothers are a common feature of the kinship system of many castes, friendly relationships generally obtain between a man and his affines. The buyer of the land, an up-and-coming local Peasant, was represented at the transaction by an over-articulate affine, his wife's sister's husband. As the transaction was about to be concluded the buyer's representative made a pompous speech in which he said that his relative did not want to be accused later of having been responsible for depriving a fellow villager of all his land. The seller should consider whether he should not retain some property in the village, be it even a manure-heap, as a symbol of his belonging to the village.

While land indeed provides the passport to membership, not everyone has the resources to buy it. And in the case of a few artisan and trading castes in Rampura, preoccupation with the hereditary calling seemed to get in the way of acquiring land. Everyone, however, needs a house, and ownership of a house was also evidence of membership. There was a hierarchy in housing, the veranda-dwellers being at the bottom and the owners of large houses with open, paved central courtyards being at the apex. The hut-dwellers were just above the veranda-dwellers in the local prestige scale. The next rung was a house with mud walls and a roof of country-made tiles. Those who owned bullocks and buffaloes tended to live in courtyard houses (*toṭṭi mane*), one part of the inner roofed portion being reserved for cattle. It was the ideal of the villager to live in a house with a big central courtyard, with many bullocks, and milch-buffaloes and cows, and many children, especially sons and grandsons. Such a household indicated that its head owned much land, and his wealth and prosperity were regarded as evidence of divine favour just as veranda-dwelling and poverty embodied the worst fears of villagers.

It is significant that the question of membership of the Harijans in the local community never came up for discussion during my stay. In this part of India, as in several others, the hereditary village servant (*chakra*) came from the Harijan caste, and the holder of the office was traditionally paid in land by the government. In Rampura in 1948 the land allotted to the original *chakra* had been partitioned among his agnatic descendants. These households formed the core of the Harijan group and additions had accrued to this group through the immigration of affines and others from neighbouring villages. The Harijans were no doubt Untouchables by caste, and as such were subjected to several disabilities stemming from the idea of pollution, but there was no doubt that they were an integral part of the village, far more so than, for instance, the Basket-makers.

One way of conceptualizing the situation obtaining in Rampura and other villages would be to regard individuals from the dominant caste as first-class members, the Harijans as third-class members and those in between as second-class members. Though such a conceptualization would be too neat and schematic, and fail to take note of the phenomenon of the 'exclusion' of each caste, including the highest, in certain contexts, it would, on the other hand, make clear the inegalitarian character of the village, the differential rights and duties attached to each category of membership, and the inclusion of everyone from the dominant caste to the Harijan in the local community.

VIII

The historic descriptions of the Indian village, first given by the British administrators early in the nineteenth century, are now seen as having been somewhat idyllic and oversimplified. Yet they have influenced the perceptions and views of generations of scholars. It is only since independence that a few social scientists, especially social anthropologists who carrried out intensive field-studies of villages, have begun critically to examine the conventional representations of the Indian village. It needs hardly be said that this is essential not only to the correct understanding of village structure and life but of the dynamics of traditional society and culture.

What I have shown in the foregoing pages is that the traditional village was far from being economically self-sufficient. Besides, the fact of its being part of a wider political system made it even less self-sufficient: the government claimed a share of the grown crop as tax, and in addition the various officials imposed their own levies on the peasant. During periods of war, able-bodied villagers were likely to be conscripted.with the result that yields declined on the farm.

Nevertheless, the village did give an impression of self-sufficiency: the villagers ate what they grew, they paid the artisan and servicing castes in grain, and a system of barter enabled grain to be used for obtaining various goods and services. There was an emphasis in the culture on getting the utmost out of the environment, every twig and leaf and the droppings of domestic animals being put to use. Castewise division of labour also added to the appearance of self-sufficiency. Moreover, the arrangements for internal or municipal government which existed in each village, and its capacity for survival in contrast to higher political entities, created the illusion of political autonomy. An incorrect understanding of the relationship between the king and the village helped to strengthen that illusion.

I have already suggested the reasons for the continuing influence of the views of the early British administrators. Social theorists such as Marx and Maine brought the Indian village into the forefront of the contemporary discussions on the evolution of property and other economic and social institutions. Marx's predominant concern with economic development the world over led him to discuss the pheno-menon of the stagnation of Indian society. He perceived the source of the stagnation to be in the isolated and economically self-sufficient character of the village community, with its impressive capacity for survival while the kingdoms which included it rose and fell. Isolation and self-sufficiency went hand in hand with the absorption of the villagers in their own tiny world and their indifference to important events occurring outside. Such villages, he held, made tyranny possi-ble, and the importance of British rule lay in the fact that for the first time in Indian history villages were undergoing fundamental changes as a result of the destruction of their self-sufficiency. Goods made in British factories, especially textiles, were displacing handlooms, and this was leading to the impoverishment of the countryside. The railways were aiding and abetting the factories in transforming village-based Indian society.

Indian nationalists used the argument of the impoverishment of the peasant and destruction of the village community to advance the cause of Indian independence. Beginning in the 1920s, Gandhi's attempt to revive the economic and social life of the village, and certain elements in his thinking such as his distrust of the power of the State, hatred of the enslaving machine, and his emphasis on self-reliance and the need for political and economic decentralization, led to a new idealization of the peasant and the village.

Perhaps it is in reaction to all this that there is now an attempt to deny that the village is a community. It is argued that the existence of caste and other inequalities make it impossible for the Indian village to be a community, for the community, it is assumed, has to be egalitarian.

But the argument that the village in India is only an architectural and demographic entity, and that it is caste that is sociologically real, does not take into account the true function of caste, which has to be viewed in the pre-British context. Given the scarcity of labour in relation to land and the resultant strong patron—client relationships, the social framework of production created bonds running counter to caste. Thus the paradox was that castewise division of labour was at the source of contracaste bonds. This was reinforced by the political system, which discouraged the formation of links across chiefdoms and kingdoms. The village was no doubt stratified along the lines of caste and land, but the productive process made it an interlocking community.

The power wielded by the dominant caste was real but it also respected certain common values, and observed the principle of 'indirect rule' *vis-a-vis* dependent castes. The leaders of the dominant caste had the maximum stake in the local community and they took the lead in all its activities. But the dependent castes also had a loyalty to the village, and were considered by the villagers themselves to have membership in it. Inclusion and exclusion operated (and continue to operate) at all levels of a caste society, and the exclusion of Harijans from certain important activities, areas, and facilities cannot therefore be interpreted as evidence of their not being part of the village community.

Finally, it must be remembered that in pre-British India there was a general acceptance of caste, and of the idiom of caste in governing relationships between individuals and between groups. Given such a framework of acceptance of hierarchy, it ought not to be difficult to conceive of communities which are non-egalitarian, their people playing interdependent roles and all of them having a common interest in survival. The argument that only 'equalitarian' societies can have local communities has to be proved, and cannot be the starting-point for evaluating hierarchical societies. Nor can an implicit assumption that 'equalitarian' communities do not have significant differences in property, income, and status be accepted as a 'sociological reality'.

REFERENCES

ASHE, GEOFFREY, 1968, *Gandhi* (Stein and Day, New York).

AVINERI. S. (ed.), 1969, *Karl Marx on Colonialism and Modernization* (Doubleday, New York).

BADEN-POWELL, B. H., 1892, *Land-Systems of British India*, vol. i (Clarendon Press, Oxford).

———, 1899, *The Origin and Growth of Village Communities in India* (Swann, Sonnenschein & Co., London).

CRANE, R. I., 1969, 'The Leadership of the Congress Party', in R.L. Park and I. Tinker (eds.), *Leadership and Political Institutions in India* (Greenwood Press, New York), pp. 169–87.

DAVIS KINGSLEY, 1951, *The Population of India and Pakistan* (Princeton University Press, New Jersey).

DUMONT, L., 1966, 'The "Village Community" from Munro to Marx', *Contributions to Indian Sociology*, 9, 67–89.

————— and D. F. POCOCK, 1957, 'Village Studies', *Contributions to Indian Sociology*, 1, 23–42.

DUTT, R. C., 1908, *Economic History of British India under Early British Rule* (Kegan Paul, Trench, Trübner, & Co., London).

Fifth Report, 1812, *Fifth Report from the Select Committee of the House of Commons on the Affairs of the East India Cy.*, 28 July 1812, 3 vols.

FRYKENBERG, R. E., 1965, *Guntur District, 1788–1848* (Clarendon Press, Oxford).

GADGIL, D. R., 1948, *The Industrial Evolution of India in Recent Times* (Oxford University Press, London).

GANDHI, M. K., 1944, *Hind Swaraj* (Navjivan Press, Ahmedabad).

GOUGH, KATHLEEN, 1955, 'The Social Structure of a Tanjore Village', in McKim Marriott (ed.), *Village India* (University of Chicago Press, Chicago, London), pp. 36–52.

HEALEY, J. M., 1968, 'Economic Overheads: Coordination and Pricing', in P. Streeton and M. Lipton (eds.), *The Crisis of Indian Planning* (Oxford University Press, London), pp. 149–72.

KARAN, P. P., 1957, 'Land Utilization and Agriculture in an Indian Village', *Land Economics*, 33, 1 (February), 53–64.

KUMAR, DHARMA, 1965, '*Land and Caste in South India* (Cambridge University Press, Cambridge).

LEWIS, OSCAR, 1951, *Group Dynamics in a North Indian Village* (Planning Commission, New Delhi).

MAHALINGAM, T. V., 1955, *South Indian Polity* (University of Madras, Madras).

MAINE, SIR HENRY SUMNER, 1890, *Village Communities in the East and West* (J. Murray, London).

—————, 1906, *Ancient Law* (J. Murray, London).

MARX, KARL, 1853, 'The East India Company—Its History and Results', *New York Daily Tribune*, 11 July 1853. Cited in S. Avineri (ed.), 1969, *Karl Marx on Colonialism and Modernization* (Doubleday, New York).

MAYER, A. C., 1960, *Caste and Kinship in Central India* (University of California Press, Berkeley, Los Angeles).

MILLER, E., 1954, 'Caste and Territory in Malabar', *American Anthropologist*, 56, 410–20.

NANAVATI, M. B. and J. J. AANJARIA, 1944, *The Indian Rural Problem* (Indian Society of Agricultural Economics, Bombay).

O'MELLEY, L. S. S., 1941, *Modern India and the West* (Oxford University Press, London).

RAMARAO, N., 1954, *Kelavu Nenapugalu* (Jeevana Prakatanālaya, Bangalore).

Report 1832: *Report from the Select Committee in the House of Commons*, Evidence III, Revenue, Appendixes (App. No. 84).

SASTRY, K. A. N., 1966, *A History of South India* (Oxford University Press, London).

SPEAR, PERCIVAL, 1951, *Twilight of the Mughals* (Cambridge University Press, Cambridge).

SRINIVAS, M. N., 1955a, 'The Social Structure of a Mysore Village', in M. N. Srinivas (ed.), *India's Villages* (Asia Publishing House, Bombay), pp. 21–35.

—————, 1955b, 'The Social System of a Mysore Village', in McKim Marriott (ed.), *Village India* (University of Chicago Press, Chicago and London), pp. 1–35.

—————, 1959, 'The Dominant Caste in Rampura', *American Anthropologist*, 61, 1–16.

————, 1962, 'Varna and Caste', in M. N. Srinivas, *Caste in Modern India and Other Essays* (Asia Publishing House, Bombay), pp. 63–9.

————, 1966, *Social Change in Modern India* (University of California Press, Berkeley, Los Angeles).

———— and A. M. SHAH, 1960, 'The Myth of Self-Sufficiency of the Indian Village', *Economic Weekly*, 10 September 1960, 1375–8.

STOKES, E., 1959, *The English Utilitarians and India* (Oxford University Press, London). Cited in Dumont 1966.

THORNER, DANIEL, 1951, 'Sir Henry Maine (1882–1888)', in H. Ausubel, J. B. Brebner and E. M. Hunt (eds.), *Some Modern Historians of Britain* (Dryden Press, New York).

————, 1966, 'Marx on India and the Asiatic Mode of Production', *Contributions to Indian Sociology*, 9, 33–66.

———— and ALICE THORNER, 1955, *Land and Labour in India* (Asia Publishing House, Bombay).

The Social System of a
Mysore Village[1]

My aim in this essay is to give a brief description of the social system of Rampura (Rāmpura), a large, nucleated village in the plains of Mysore District in Mysore State in South India. Rampura is a village of many castes, yet it is also a well-defined structural entity. Comparisons between Rampura and other villages may throw some light on India's rural social systems generally.

With its 1,523 residents in 1948, Rampura must be ranked among the largest 5 per cent of the villages in Mysore State. Seventy-seven per cent of all villages in the state contained less than 500 persons in 1941, while 93 per cent contained less than 1,000 persons (*Census of India*, 1942, p. 12).

Structurally, also Rampura is much more complex than the average village, for it contains nineteen Hindu castes and Muslims. Each of the four largest castes of Rampura has a strength of more than one hundred persons, and together these four number 1,274 out of the total village population of 1,523. Others are much smaller. Below is a list of all the castes of Rampura, arranged in order of their population strength in the village, together with the traditional calling of each caste (Table 1).

Traditional callings are implied in the vernacular names of most of the castes. For convenience in description, I refer to them by English terms which are equivalent to their vernacular names—'Potter' for Kumbāra, 'Peasant' for Okkaliga, etc. Exceptions are, however, made for the religious terms 'Muslim' (Musalmān), 'Brahman' (Brāhmana), and 'Lingāyat', and for the 'Untouchable' Holeya.

[1] I have spent twelve months doing fieldwork in Rampura. An initial stay of ten months in 1948 was made possible by the generosity of the University of Oxford. A second visit of two months in 1952 was financed by the M. S. University of Baroda. I wish to thank the University of Manchester for the award of a Simon Senior Research Fellowship which permitted me to spend nine months during 1953–4 on the analysis of field data.

The Castes and Their Occupations

Each caste in Rampura is traditionally associated with the practice of a particular occupation. This does not mean, however, that all the members of a caste or even a majority of them do in fact always follow their traditional calling. And even when they do follow a traditional calling, they need not do so to the exclusion of another calling. In fact, some nontraditional calling may be economically more remunerative than the traditional one.

TABLE 1

POPULATION AND TRADITIONAL CALLINGS OF CASTES IN RAMPURA (1948)

Name of Caste	Traditional Calling	Population
Okkaliga	Peasant	735
Kuruba	Shepherd	235
Musalmān	Artisan and Trader	179
Holeya	Servant and Labourer	125
Gāniga	Oilman	37
Acāri—Kulācāri and Matācāri	Smith	35
Lingāyat	Non-Brahman Priest	33
Ediga	Toddyman	24
Kumbāra	Potter	23
Banajiga	Trader	22
Kelasi	Barber	20
Besta	Fisherman	14
Korama	Swineherd	12
Agasa	Washerman	7
Mēda	Basket-maker	7
Brāhmana—Hoysala Karnātaka	Priest and Scholar	6
Brāhmana—Mādhva	Priest and Scholar	6
Brāhmana—Śrī Vaisnava	Priest and Scholar	3

Older and more conservative persons in each caste tend to regard the traditional calling as the proper one. Each takes pride in the skills which are required for his traditional calling and regards these skills as natural monopolies of his caste. For instance, Brahmans are assumed not to possess agricultural skill. If, in fact, a Brahmin villager does show some skill in agriculture, then other villagers may express their surprise. The Brahman priest of the Rāma temple in Rampura shows such skill and is frequently to be seen carrying a basket and a sickle to the fields. His agricultural skill is admired if not envied, but he is also criticized for alleged neglect of priestly duties in favour of agriculture. On the other hand, the Peasants of nearby Bella village are criticized because of their urban ways and their lack of skill in rice cultivation, to which they are comparative newcomers.

Contrary to popular impression, the traditional calling is not unchangeable. Changes are especially common at the present day when members of all except the lowest caste are seen opening shops and starting rice mills and bus lines. In the recent past, too, castes have changed their occupations, and changes probably took place in the remoter past as well. For instance, the Shepherds of Rampura have changed over from keeping sheep and weaving blankets to farming. In Rampura none of the Fishermen fishes for a living, and in the neighbouring village of Kere most of them have taken to agriculture. Only two families of Oilmen still work the traditional bullock-powered mills to extract oil out of oilseeds. The rest of the Oilmen are engaged in petty trade or work in agriculture as tenants, servants, or labourers. During my stay in the village I came across a party of itinerant Washermen who had discarded their traditional occupation in favour of digging wells.

Again, a caste may have more than one traditional occupation. For instance, the Toddyman not only taps toddy from palm trees but also sells it. The Oilman extracts oil from oilseeds and sells oil, oilsoaked cotton torches, and edibles fried in oil. The Fisherman not only catches fish but also sells them. The Shepherds used to keep sheep, make blankets, and sell both sheep and blankets. In brief, division of labour is not highly developed: the specialized caste often performs many operations to produce its special articles and then markets them as well. Furthermore, in addition to their separate traditional occupations, all castes down to the Untouchable have for long commonly practiced agriculture. Even the Brahmans have done so, although there is a scriptural ban against their doing so (Manu X, 84).

The extent of conformity, change, flexibility, and overlapping in the traditional occupations of the castes of Rampura may best be seen from a caste-by-caste review of the actual situation.

Peasants and Agriculture

The Peasant caste is the only caste in the area of Rampura which has agriculture as its sole traditional occupation. Peasants are the most numerous caste in Rampura village, numbering as they do nearly half the population. They tend to predominate generally in the districts of Mandya and Mysore. The bulk of the Peasants are actually occupied on the land, either as owners, as tenants, as labourers, or as servants. All the biggest landowners in Rampura are Peasants.

But it may be repeated here that most other people in the village, regardless of their differing traditional occupations, are also engaged in agriculture in one capacity or another. Brahmans were found along with Peasants among the biggest landowners of the area until about thirty years ago. In Rampura at present, only the Traders, Basketmakers, Washermen, Swineherds, and a few Fishermen do not engage in agriculture in any way. The total number of the nonagricultural population in the village does not exceed 100 persons. According to the Harvest Scheme List for 1946–7, more than one-half of the families of Rampura required no grain ration cards. Forty-nine families were reported to be growing enough grain for their own consumption, while 100 families were reported as producing a surplus for government purchase.

The remaining agricultural families—132 out of the total number of 281 families—were reported to be marginal agriculturists who did not grow enough grain to subsist without the help of a government ration. If any bias exists in these figures, it is probably in the direction of minimizing agricultural production so as to escape selling grain to the government. Agricultural work and income are a substantial element in most villagers' lives, even if many participate in agriculture only as servants or as casual, seasonal labourers.

Like the Brahmans before them, some Peasants and members of other, older, landed castes have been attracted to the city and have given up their land. Rampura boasts of three Peasant and one Lingāyat college graduates, three of whom are employed in the government while one works as a lawyer in a nearby town. Between 1948 and 1952 two Peasants started rice mills, and another started two bus services to Mysore. The bus-owner has bought a house in Mysore and has also built a few houses for renting in Bella. Some Peasants are engaged in trade: they keep teashops, sell groceries and cloth, and hire out cycles. A few Peasant youths show a keen awareness of the changed political situation and have ambitions of capturing political power.

Priestly Castes

Brahmans and Lingāyats are the traditional priestly castes of Rampura. This does not mean that every Brahman or every Lingāyat in the village is actually a priest. In fact, the bulk of Brahmans and Lingāyats are engaged primarily in secular occupations while even those who practice priesthood often also engage in subsidiary occupations such as agriculture and moneylending.

To say that Brahmans and Lingāyats are priestly castes means only that some individuals from these castes serve as domestic or temple priests. Brahman and Lingāyat priesthoods are somewhat distinct from one another. Lingāyats are temple priests only at temples where the god Siva in one of his numerous manifestations is worshiped. Brahmins do not in theory enter tempels where Lingāyats are priests; Lingāyats do not enter temples where Brahmans are priests; but members of other Non-Brahman castes may enter temples where either Brahmans or Lingāyats are priests. There is a distinction also in domestic priesthood: Lingāyat priests are called to officiate at births, weddings, or other ritual occasions only in the houses of other Lingāyats, while Brahman domestic priests may be called by any of the other high Non-Brahman castes.

In addition to Brahman and Lingāyats priests, there are priests to be found in every other caste. Such priesthood tends to run in certain families, one of the sons being initiated into the priesthood on the death of the father or paternal uncle. Practicing the priesthood might mean conducting regular, i.e. daily or weekly, worship in a temple, or offering worship on certain special occasions. In Rampura, for instance, the biggest lineage among the Peasants has a small temple in which are housed two goddesses who were brought into the village by the founders of the lineage. Two men of the lineage are priests (*guḍḍa*) and offer worship at the temple once a week, besides officiating at the periodical festivals held in the deities' honour. This lineage also supplies a priest to the Māri temple in the village. Māri is the goddess presiding over plague, smallpox, and cholera, and her propitiation protects the village from the dreaded epidemics. The village as a whole is interested in warding off epidemics. Brahmans residing in the village make offerings to Māri when someone in the house is suffering from plague or cholera, or when an epidemic is about. The Untouchables have another separate shrine to Māri and have recently established a separate shrine for worshiping the deity Rāma. Only Untouchables worship at these shrines. In brief, it is wrong to assume that priests are always recruited from the Brahman and Lingāyat castes and not from other castes.

Nor are Brahmans of Rampura by any means restricted to practicing some sort of priesthood to earn a living. At the beginning of 1948, one of the three Brahman families of Rampura was the immigrant Śrī Vaiṣṇava family of the village doctor. A second was the old native family of the village postmaster, who owned a little land and did contract work on the canal in summer for the government. The third family was that of the priest who had been called by the local elders over twenty-five years ago to come and officiate at the Rāma temple. This temple priest gave part of his time to temple work and part to cultivating the lands with which the temple was endowed. The priest also occasionally officiated at a wedding or other ritual occasion in the village if he was invited to do so. But he was really not entitled to do this, as the right belonged to another Brahman who was residing in Hogur, a large village about five miles from Rampura. The joint family of this absent Brahman had the right to provide a priest at weddings or other ritual occasions among all high castes, barring Lingāyats in the sixty-five villages forming part of Hogur Hobli. This right is enforceable. Any other Brahman acting in his place without his prior consent might be asked to explain his conduct before the village panchayat. The man employing him would also be liable. In such a case, the panchayat would fine the guilty parties. Such a right would be enforceable also before a government court of law.

In 1948 this domestic priest was acting also as the accountant (*śānabhōg*) of Rampura and of the adjacent village of Gudi in place of the true holder (*barābardār*) of that accountantship. In 1949 a descendant of the joint family of the true holder was able to assert his claim to the office. His ancestors had lived in Rampura and had owned a considerable quantity of land there. But, like many Brahman families in this area, they had left the village for the city to secure Western education and urban jobs. The family had come down in the world, and the return of one of its members to Rampura to take up an extremely ill-paid accountantship was a measure of its fall. The family sold all its land, most of which was bought by rising Peasant families—a familiar sequence of events in the neighbourhood. The postmaster, too had spent some time in Mysore in his youth, and later returned to Rampura only because he could not get a job in the city.

Like the Brahmans, the Lingāyats of Rampura practice either priestly or secular occupations according to their circumstances. There are two priestly lineages among the local Lingāyats. One of them, which enjoys the lucrative priesthood of the Mādēśvara temple at Gudi, which is about a mile away from Rampura, in split into two joint families, each of which performs the tasks of priesthood in alternate years. One of the two joint families further split up into four families during the latter half of 1948. Each of the four brothers now acts as priest once every eight years. A second Lingāyats lineage furnishes a priest for the temple of Basava, the bull on which the deity Śiva rides. The priest cultivates the temple's endowed lands. The other Lingāyats in the village are engaged in agriculture and in trade.

Serving Castes paid at the Harvest

Five castes—the two Smith castes, Potter, Washerman, and Barber—serve the cultivators regularly and are paid a certain quantity of grain at harvest time by those whom they have served during the year.

'Smith' is a blanket term which includes two castes and three groups of occupations in Rampura—working with wood and iron, building houses, and working with precious metals and stones. Generally speaking, the last-mentioned belongs to the Matācāri subdivision, and the first two to the Kulācāri sub-division. There is, however, one Matācāri Smith who works with wood and iron and one Kulācāri Smith who works with wood and iron as well as with precious stones. One Smith who works with wood and iron also occasionally helps a housebuilding Smith in his occupation. The Matācāri drink alcoholic beverages while the Kulācāri do not. Each is a distinct endogamous group.

The hereditary Potter of Rampura lives in Gudi. He makes pots and pans and supplies them to several families on occasions such as birth, marriage, and death. He does not seem, however, to make tiles for the inhabitants of Rampura. There are, in addition, a few Potter families resident in Rampura, only one of which (composed of a man, his wife, and an immigrant assistant) carries on the traditional occupation. The head of this house owns a little land which he personally cultivates, and during the nonagricultural season he makes pots, pans, and tiles. He makes them, however, not for grain payments but for cash.

The Potters' trade hangs on old tastes. Their traditional products are favoured by the bulk of the villagers, who believe that food cooked in earthen pots is 'cooling', unlike food cooked in metal vessels, which is 'hot'. Even in the rich headman's house *rāgi* flour is still cooked in huge earthen pots. But Brahmans and some of the richer persons in other castes now use metal vessels, usually of brass or copper, for the use of metal vessels is thought to confer more prestige on their owners than the use of earthen ones. As the supply of cash in Rampura has increased in the last fifteen years or more, there seems to have been an increasing tendency among the poor to buy metal vessels. Similarly, houses roofed with the traditional Potters' tiles are believed to be cooler in summer than those roofed with factory-made tiles, yet there is a tendency for the richer and more urbanized villagers to use factory-made tiles for prestige reasons. In 1948 three buildings in Rampura had been roofed with factory-made tiles. If this tendency increases, the Potters may be forced in course of time to give up their traditional calling.

While Smiths and Potters serve everyone in the village irrespective of caste, Barbers and Washermen serve only those castes which do not pollute them by contact. A Smith is not pulluted by handling an Untouchable's plough, nor is a Potter polluted by giving a pot or a few tiles to an Untouchable. But physical contact with the customer is implied in the services which are rendered by the Barber and Washerman. Swineherds and Untouchables are therefore excluded from the services of the Barber and Washerman castes: they have to provide these services for themselves from within their own respective castes.

The Barber of Rampura shaves his customers once a week or once a fortnight, depending on the amount of grain paid. He does not shave his customers on certain inauspicious days, such as the days of the new and full moon. The Barber provides special services on ritual occasions, such as birth, death and wedding, in return for extra gifts in cash and kind.

Two families of Washermen serve Rampura. One of them which is resident in the neighbouring village of Bihalli washes the clothes of only a few Rampura families, notably those of the headman and Barbers. Betweem 1948 and 1952, however, the Bihalli Washerman lost the custom of the Rampura headman. The headman's reversion to the local Washerman conforms with the common tendency for ties with servicing castes to be confined within the same village. The Barbers' patronage of the Bihalli Washerman is due to a long-standing dispute between themselves and the Washermen of Rampura.

The Washerman washes clothes, returning the clean garments to his customers once a fortnight. Between his visits, his customers wash their own clothes, since it is only the better-off Peasants who own more than one change of clothes. The washing of their customers' menstrual clothes, which women of the Washerman caste do, is regarded as degrading and defiling. Men of the Washerman caste wash only their male customers' clothes.

Other Castes

There are nine resident castes which do not provide regular services or receive grain payments at the harvest: Shepherds, Oilmen, Toddymen, Fishermen, Basket-makers, Swineherds, Traders, Untouchables, and Muslims. And there are several other nonresident castes which also enter into the economy of the village.

None of the Shepherds now keeps sheep, but a few of them still make coarse blankets from wool. They obtain this wool either from the few local Peasants who have kept small flocks or from neighbouring villages. The wool is 'paid for' in blankets, there being a recognized rate of exchange. There is very little pasture land in or around Rampura, and this is one of the reasons why local Shepherds have had to change to agricultural occupations.

Two Oilman families, the heads of which are brothers, carry on the traditional calling by extracting oil from seeds. They also own some land, which they cultivate. In addition, they go periodically to the great temple of Mādēśvara in Kollegal Taluk in Madras State, where they sell torches made of oil-soaked rags tied to lengths of bamboo. Pilgrims go round the temple with torches in their hands. Selling torches is apparently a profitable business: the brothers have bought some riceland, and have saved some money in addition. The other Oilmen in Rampura are mere labourers, servants, or petty traders.

One large joint family of Toddymen owns some riceland and a cloth shop. The head of this joint family also sews on a sewing machine in his shop. He seems, furthermore, to have an interest in a toddy shop in Hogur. A second Toddyman family in the village makes and sells mats made out of toddy palm leaves—a traditional occupation of women of the caste. But some members of this matmaking family also engage in odd coolie work. The toddy shop outside Rampura itself is run by a Toddyman widow. She is the only member of local caste group who is entirely dependent on this traditional calling, a calling which is considered to be low.

Three of the four Fisherman families in the village live by coolie labour, while the head of the fourth operates a sewing machine.

The Basket-makers are recent immigrants, and they make baskets, screens of split bamboo, winnowing fans, hencoops, etc., for sale either locally or in a nearby weekly market (*sante*). They buy the bomboos in Mysore. Their contact with towns is greater than that of the other castes. They are not regarded as belonging to the village, as every few years one batch of Basket-makers is replaced by another.

The men of the Swineherd caste herd the swine which they own, and the oldest woman among them goes about villages telling fortunes. Both are traditional occupations. The Swineherds speak among themselves a dialect of Telugu, and have a culture which is somewhat different from that of the other castes in the village. They live in huts on the outskirts of the village. From the beginning of the rainy season until the end of paddy harvest they live in the village headman's mango grove to the north of the village; then they spend the summer in the headman's paddy field to the southeast of the village. Their transhumance is due to the headman's desire to have his fields fertilized with pig manure.

There are three joint families of Traders, all of which are engaged in trade. Groceries and fried eatables are sold in their shops. One Trader has also kept a cloth store and a sewing machine which he operates himself. Two other Trader youths are able to sew, and one of them left Rampura some time ago to settle down in Harigolu, his wife's village. It is interesting to note that none of the Traders owns land.

Of thirty Untouchable families, fifteen are cultivators, and fifteen live by coolie work. Some of the cultivators are *cākaras*, or hereditary village servants, whose duty it is to assist the headman and accountant in the collection of land tax. The men of the families which live by coolie work are either agricultural labourers or servants. Most Untouchable women do coolie work. They transplant paddy shoots, weed, help with harvest work, trim the acacia trees during summer, clean grain, grind flour, etc.

In 1948 there were thirty-nine Muslim families in Rampura of which thirteen owned or cultivated land, fifteen worked as labourers, and seven were engaged in petty trade. The occupations of butcher, shoesmith, tinker, doctor of Unani (Muslim) medicine, and plasterer were each performed by a Muslim. There were, in addition, two Muslim tailors. Some show much enterprise in their commercial activities: though Hindus own the mango groves in this area, the entire trade in mangoes is in the hands of Muslims. The poorest Muslims act as middlemen, borrowing money for short periods at high rates of interest and working on very small margins; it is not surprising that their enterprises collapse occasionally. The Muslims are therefore occupationally as well as spatially mobile. The bulk of the Muslims of Rampura are recent immigrants and are not yet regarded as fully belonging to the village.

Several itinerant castes also visit Rampura and neighbouring villages, commonly in the summer. Sawyers from the lowlands of the Tamil country come to saw timber for building. Tilers, Hunters (Bēḍas) Well-diggers, and a few castes of entertainers also come— Gāruḍiga magicians, Muslim snake charmers, Pipers who play pipes through their nostrils, etc. Shepherds from villages very near Mysore City come to Rampura if a long drought has burned up the grass in their area. They take their flocks along the banks of the Cauvery River. There is customary arrangement by which the Shepherds stand their flocks so as to manure a man's field at night and receive in return a meal or the raw materials of a meal. Most peripatetie castes wander about only after the agricultural season is over. During the agricultural season they stay at home to raise crops.

The Traditional Economy of Land and Grain

In the traditional economy of this area, money seems to have played a minimal part. Even at the beginning of this century cash was scarce, and the buying power of a rupee was much greater than it is today. Barter still prevails in Rampura, and it was much more widespread in the past. Today, as long ago, a farmer's wife barters paddy for dried fish, vegetables, and betel leaves. Fruit-sellers are frequently paid in paddy. Though the farmer grows paddy, his staple is *rāgi* (finger millet). Before World War II it was not uncommon for villagers of Rampura to drive carts of paddy to Hunsur in the west in order to exchange the paddy which they had grown for the *rāgi* which they would eat.

Within the village of Rampura the usual way of paying for services was and is in grain, or in land, the source of grain. The various kinds of payment may be arranged in a hierarchy of prestige, with payment in land at the top. Land is the most permanent form of payment. A piece of land may be attached to an office, as, for example, to the office of a village servant or of a temple priest; or it may be given to someone for rendering a service, as it has occasionally been given to a servant who had faithfully served his master's family for a long time. The implication of payment in land is that the land is to be held so long as the office is held or the service performed. Prestige is also attached to grain payments, though their prestige is less than the prestige which is attached to payments in land. The prestige of grain payments is understandable, for the ability to pay in grain is the result of rights in land, rights either of ownership, or of tenancy. Grain payments also imply enduring relationships, and enduring relationships are valued. The payment of a crop growing on a piece of alnd is regarded as intermediate in prestige between payments in land and payments in grain. Finally, at the bottom of the hierarchy of prestige are payments in cash.

The principal temples in the village—the temples of Rāma, Basava Haṭṭi Māri and Kabbāḷa Durgada Māri—are endowed with agricultural land. Excepting the lands of the Hāṭṭi Māri temple, all the endowment lands are ricelands. The priests enjoy the fruits of these lands, and the priests of the Rāma, Basava, and Mādēsvara temples claim, in addition, a headload of paddy with straw from everyone who grows enough to have some to spare.

In the past, three Brahman families, including the hereditary domestic priest of the village, held grants of riceland. The reason for the grant to the priest is clear; the reasons for grants to the others are less clear, though and gift to poor Brahmans has always been regarded as a pious act. I have not made a study of these grants, but I would assume in all these cases that the state was the donor. Where the gift is attached to a temple, or to an office, such as that of a village domestic priest or servant, enjoyment of the land is conditional on performance of the duties of the office. But after the lapse of a generation or two such property tends to be treated as the private property of the donee. The village priest of Rampura, for instance, sold a portion of his land a few months before I started work in the village. And it is very common, if not universal, for such property to be divided, like any other property of a joint family, among the heirs of the deceased person.

In this area it is common for the village servants, like the village priests, to be paid for their services in the form of land. In Rampura these village servants are Untouchables. They are required to assist the headman and the accountant in the discharge of their duties. The land of the village servants, too, has been divided, as if it were joint family property.

There is, however, a difference in responsibility as between priests, on the one hand, and village servants, on the other. Village servants, like the headman and the accountant, are servants of the state government. They are subject to the authority of the revenue officials and, in case of extreme incompetence or corruption, are liable to dismissal by the government. But the priests are responsible only to the village community represented by the elders. If a priest enjoys the fruit of his lands without performing the duties of his office, he may be controlled only by public opinion and by religious sanctions.

The village headman (*paṭēl*) and accountant are paid indirectly from the land. They keep for themselves a cash commission figured as a percentage of the land taxes which they collect. This commission form of payment may have been instituted to encourage them to collect the full tax from the cultivators.

Grain payments, which stand below land payments in point of prestige, are made for rent, for regular services, for charity, and for labour under certain circumstances. Tenants in Rampura universally pay rent (*guttige*) to the landowners in the form of grain. Such a grain payment may be either a fixed share of the harvest—one-half or one-third—or a fixed amount of paddy. Rents vary at present from four to six *khaṇḍis* of paddy per acre, one *khaṇḍi* being equal to 180 *seers* (about 360 pounds).

The Smith, Potter, Washerman, and Barber are paid a fixed quantity of grain, called *aḍade*, during the paddy harvest of each of their more substantial customers. These four are entitled to receive also small quantities of the pulses, which are harvested in late summer, of the vegetables and chilies grown in the vegetable plot in the paddy land, and a few cubes of jaggery, if jaggery is being made. They are given additional gifts of grain, money, and food for their services on special ritual occasions. When a Washerman or Barber refers to a family as his '*aḍade kuḷa*', i.e. 'grain-payment family', he implies, first, that the family is wealthy enough to pay annually in grain and, second, that the relationship between them is an enduring one. Such grain payments are made only by those farming families which grow a surplus of grain. Cash is paid for every act of service by those who do not grow a surplus. The quality of the service rendered by the Smith, Potter, Washerman, and Barber depends on whether the customer pays annually in grain or not, on the quantity of grain paid, and on the customer's general social position.

There are minute regulations governing the conditions of service of these grain-paid castes. The quantity of grain paid to the Smith varies from family to family but varies always at a fixed rate according to the number of ploughs owned. The Smith is not required to make the entire plough anew; he only beats into a ploughshare and sharpens the piece of iron brought by the customer. The customer must first buy the piece of iron, either from a local trader, in a weekly market, or in a nearby town. He must take the iron along with a length of acacia wood to the Smith. The customer has in addition to contribute his own labour. The chips from the wood are the Smith's perquisite. The Smith repairs all the agricultural implements as part of his retaining fee, but he has to be paid separately for repairing carts.

While the amount of grain paid to the Smith depends on the number of plough, the Barber's depends on the number of adult males in the houses he serves. The question as to when for purposes of grain payment a boy becomes an adult male is a matter which must be argued between the Barber and the head of a household. The amounts of grain paid to the Washerman and Barber depend upon the total number of adults in the household. Women mean more work for the Washerman, as a woman's saree is a longer garment than a man's lower cloth. Normally, the Washerman and Barbers would exchange their services without pay, but since a dispute has arisen between them, the Barbers have been getting their clothes washed by the Bihalli Washerman, while the Rampura Washerman pays the Barber for his shave.

Next in order of importance after the payments to the servicing castes are the payments of headloads of paddy with straw which are given to the priests of the principal temples in the village. These are said to be contributions towards the daily offerings of cooked rice which are made to the deities.

Headloads of paddy with straw are also given to mendicants, agricultural servants, and the importunate poor. The giving of grain at harvest time to the poor and to mendicants is an act of piety. The performance of such a charitable act results in good, both materially and spiritually, here and hereafter.

The richer landowners in Rampura sometimes pay their servants partly in cash and partly in grain. The servant is given the paddy crop growing in three or four plots. He is then required to supply manure to these plots and to do the weeding and harvesting himself. A rich landowner adopts this mode of payment when he wants to be certain of the supply of labour. Such payment is more highly valued than money, as the food-buying power of money varies. Paddy is preferred because it is food, although it may be converted into money if necessary.

Labour is especially scarce during the paddy transplantation season (July–August). At that time the labour of many people, men and women, has to be concentrated in a particular field for a particular day, or two, or three. Villagers are fond of saying that transplantation (*nāti*) is like a wedding: by this they mean that it is a collective undertaking which has to be finished within a brief and specified period of time.

Labourers who help at transplantation are usually paid in cash, since grain is scarce at transplantation time. Only when both the landowner and the labourer belong to the same village, and when the landowner is known to be a reliable man, does a labourer agree to be paid in grain at the harvest for the work which he has done at transplantation time. Wages for labour during the harvest are paid in paddy.

Payments in the form of land and grain bring home to everyone in Rampura the interdependence of the castes. The Smith, Potter, Washerman, and Barber, the priests of the principal temples in the village, the labourers who helped during transplantation and harvest, other servants, poor friends, the village sweeper, the butcher—all these and many others may be given gifts of headloads of paddy with straw. In such a traditional economy the divisiveness of caste endogamy and the barriers against commensality and free contact are only a part of the story. Division also implies interdependence. Each caste is aware that it is not self-sufficient. Payments in land and grain may be said to dramatize this fact.

While the traditional economy of Rampura may be described as one of land and grain, a land tax in the form of cash seems to have been paid at least since the time of Chikka Deva Raja Wodeyar (AD 1628–1704). The role of money has been increasing, however, in recent times. The prevalence of high prices for food grains during World War II brought large sums of money into Rampura and other villages, especially into the hands of the larger landowners. Some portion of this money was invested in new enterprises such as rice mills, buses, and urban houses for rent, rather than in older forms of investment such as land, usury, rural houses, and women's jewelry.

Sources of Flexibility

Under the caste system the nonagricultural castes are assured not only of a monopoly over their traditional callings but also of the freedom to choose among certain alternatives—especially agriculture and trade on the external market—alternatives which give flexibility to the traditional social system, yet help to preserve its form.

The stability of caste monopolies is enforced by family inheritance. That is, the right to serve a particular family—the right of making ploughs for it, or of periodically shaving the heads of its male members, or of washing its clothes—is treated as a heritable and divisible right. Thus the partitioned brothers of a Barber family divide among themselves the families which they were all jointly serving before partition. On the other hand, all the partitioned sons or brothers of a patron family continue to patronize the same Barber, Smith, Potter, and Washerman who used to serve them before partition. A respect for enduring relationships is to be seen everywhere.

But this tendency toward stability does not mean that continued unsatisfactory behaviour on either side will be tolerated. After protesting to the village elders, the aggrieved party will break off the old relationship and form a new relationship with another. Shifting relationships may ultimately make one Smith or Potter more popular and therefore richer than others. Such shifting of relationships is also partly responsible for the rivalry which exists between members of the same nonagricultural caste in a village. Each Smith has one eye on the customers of another Smith. Occasionally the elders of the village are called upon to settle a dispute between two Smiths, one of whom alleges that the other is trying to poach on his circle of customers. Such a dispute highlights a feature of the caste system that is normally in abeyance: while there are undoubtedly strong ties binding together the families of any non-agricultural caste in a local area, strong rivalry often also exists among them (cf. Gough 1952, p. 535). The kinship links, agnatic or affinal, which prevail among them act both as checks and as stimulants to this rivalry.

Rivalry within the caste tends also to encourage the formation of friendships outside the caste (cf. Marriott 1952, p. 873). Thus the division into castes brings together the various castes in a village or local area by means of two linked processes: first, the existence of occupational specialization brings the different castes together, and, second, the rivalry for customers splits the members of a caste and forces them to seek friends outside.

Occupational specialization has its limits, however, since no single village or group of a few neighbouring villages can support an indefinite number of Smiths, Barbers, Potters, Washermen, Oilmen,

Fishermen, Shepherds, Traders, Basket-makers, Swineherds, or Priests. It is obvious that the bulk of the people of a village, who live by agriculture, have to 'carry', as it were, the nonagricultural castes by their payments of grain and of grain-producing land.

When there are too many persons in a nonagricultural caste group, the excess may migrate to a nearby town. They may stay there practicing their traditional occupation or a new one. Change of occupation is more likely if the town is a Wester-type town rather than a traditional one. Movement to a town, if it is also followed by a change of occupation, may lead to the formation of a new caste, and eventually to change of caste rank. Migration to another village may also occur, especially when the migrant has relatives there or a patron who can sponsor him. It is, however, theoretically possible for a nonagricultural caste to live largely independent of the village economy by producing for a weekly market. Thus Basket-makers, Oilmen, Potters, and others might make a living largely outside the village economy. This is, however, infrequent, as a weekly market does not provide the same security of livelihood that grain payments do, unless the market is an exceptionally big one. It is more usual for an artisan to trade at a weekly market simply to supplement his more certain income within the village.

Another alternative for the surplus nonagriculturists is to abandon their traditional callings and to take up agriculture as tenants, as servants, or as labourers.

The rule permitting all castes to take up the common occupation of agriculture keeps the caste system going by drawing off the surplus persons in the nonagricultural castes. If it has not been for the alternative of agriculture, the occupational aspect of the caste system, and with it perhaps the entire caste system, would have broken down under the great increase of population which has occurred during the last hundred years.

Agriculture gives flexibility to the social system in yet another way. In the traditional rural economy, ownership of land is the most important source of wealth, and is the means by which individuals lift themselves up in the local prestige system. If a member of a low caste becomes rich, he invests a good part of his wealth in land. There is, in such a case, an inconsistency between his caste rank and his wealth. This is seen, for instance, in the case of one Toddyman of Rampura, who is better off than many members of higher castes. For purposes of contribution to common village festivals he is put in the second division along with some members of higher castes. The lower divisions also include some who belong to higher castes. Brahmans are in a sense removed from comparison with wealthy members of lower castes, for Brahmans do not contribute to festivals, nor do Muslims or Untouchables.

I suggest that here again is a situation which makes possible both the formation of a new caste and the upward movement of a caste in the hierarchy. When a member of a low caste owns some land, there is a tendency on his part to Sanskritize his ways and customs (cf. Srinivas 1952, pp. 30–1 and *passim*). In the past, a caste's claim to high position went hand in hand with Sanskritization. Nowadays, members of low castes who hold official positions also show a tendency to Sanskritize their ways.

Traditional Political Organization : The Dominant Caste

The existence of caste courts has been interpreted as proof of the strength, if not of the autonomy, of a caste. But the political strength of castes, like their occupational specialization, is only a part of the story. The settlement of disputes in the village occasionally brings out the importance of one caste which is locally dominant, and the dependence of the other castes on it. The concept of the dominant caste is important for understanding intercaste relations in any local area, and for understanding the unity of the village.

A caste may be said to be 'dominant' when it preponderates numerically over the other castes, and when it also wields preponderant economic and political power. A large and powerful caste group can more easily be dominant if its position in the local caste hierarchy is not too low.

The elders who govern Rampura owe their power not to legal rights derived from the state but to the dominant local position of their Peasant caste group. Their power is so great that it is not unknown for cases pending before the state official courts to be withdrawn in order to be submitted to their adjudication.

Justice can be swift and cheap in the village, besides also being a justice which is understood as such by the litigants. The litigants either speak for themselves or ask a clever relative or friend to speak on their behalf. There are no hired lawyers arguing in a strange tongue, as in the awe-inspiring atmosphere of the urban state courts. I do not hold that the justice administered by the elders of the dominant caste in Rampura is always or even usually more just than the justice administered by the judges in urban law courts, but only that it is much better understood by the litigants.

The elders of the dominant Peasant caste in Rampura administer justice not only to members of their own caste group but also to all persons of other castes who seek their intervention. Even now, in the rural areas, taking disputes to the local elders is considered to be better than taking them to the urban law courts. Disapproval attaches to the man who goes to the city for justice. Such a man is thought to be flouting the authority of the elders and therefore acting against the solidarity of the village.

The elders of the dominant caste are able to dispence justice to everyone because, where necessary, they apply the code which the disputants recognize and not the code of their own caste. They may regard their own caste code as superior, but they recognize that members of other castes have a right to be governed by their own codes.

The minority castes in Rampura, including the Muslims, seem only too ready to take their disputes to elders of the Peasant caste for settlement. The sentiment that disputes should be settled within the caste does not seem to be very strong. There is, on the other hand, a tendency for the poorer people to take their cases, even quarrels within the joint family, to their patrons, who are usually Peasants. Peasant elders may be called upon to decide cases in which all the litigants are Brahmans, or Untouchables, or Muslims. In one case which I collected from the village of Kere the Peasant elders on appeal set aside a decision given by the elders of the local Fisherman caste to some of their own castemen on the grounds that the decision was unjust and motivated by malice.

Sometimes, however, the elders of the dominant caste either give permission, or actually suggest, that a case be referred to the caste court of the disputants' own caste. They may do so when an intricate point of caste custom is at issue, or when the witnesses are spread over several different villages. Other considerations may also prevail, such as the question of jurisdiction. A gereral rule is difficult to state, because there is variation from village to village and from caste to caste. On the whole, the elders of the dominant caste show respect for the customs of each minority caste and for its elders, and vice versa.

The elders of the dominant caste are spokesmen for the village. Trouble would ensue for a person who did not show them proper respect. They are able to supply or to withhold information about people living in the area. Their co-operation may be essential for rendering effective a sentence passed against an individual in a caste court. Their friendship may be needed in some future transaction in land or cattle, or in soliciting a loan from someone, or in finding a bride.

By comparison with taking a dispute to the village elders, taking a dispute to the caste court is a procedure not unattended by an element of risk. A man can be certain of receiving consideration, if not kindness, at the hands of the elders of his own village; he cannot be as certain of it at the hands of his caste elders, some of whom belong to different villages and some of whom he does not know well. A caste court is not unlikely to decide a case entirely on a point of law—a thing which is less likely to happen in a court of village elders who are well acquainted both with the persons and with the circumstances of each case.

The Caste Hierarchy

The essence of hierarchy is the absence of equality among the units which form the whole: in this sense, the various castes in Rampura do form a hierarchy. The caste units are separated by endogamy and commensality, and they are associated with ranked differences of diet and occupation. Yet it is difficult, if not impossible, to determine the exact, or even the approximate, place of each caste in the hierarchical system.

Before the castes can be ranked as unequal, they must, of course, exist as separate units. Separation of the castes is achieved, first, through endogamy. The effects of caste endogamy are, on the one hand, to deny a powerful potential means of forging solidarity among different castes and, on the other hand, to increase solidarity within each caste.

Separation of the castes is achieved, second, through restrictions on commensality. Complete commensality may be said to exist only when all persons, men as well as women, accept cooked food and drinking water from each other. Thanks to the pervasive concept of pollution, each person accepts drinking water and cooked food only from castes which he regards as equal or superior to his own. Acceptance from an inferior caste would pollute him and would entail his performing, among other things, a purificatory ceremony to regain his normal ritual status. The pollution conveyed by contact with lower castes is one of several kinds of pollution (cf. Srinivas 1952, chaps, ii, iv).

Women are more particular than men about commensal restrictions. For instance, while the men of equal castes may eat food cooked by each other, women do not do so. Thus the Peasant and Shepherd men eat food cooked by each other, but their women do not. Complete commensality prevails only within a single caste.

There is a hierarchy in diet and occupation to which the caste hierarchy is related. Vegetable food is superior to meat, and there is again a hierarachy in meat. Beef is the lowest of all, while pork, chicken, and mutton follow in order of superiority. Cattle are sacred to all Hindus, and no one kills a live cow or bull for food Only dead cattle are eaten by Untouchables. Thus beef-eaters are also eaters of carrion. The domesticated pig goes about the village eating, among other things, human ordure, and this is why eating the pig is considered a mark of very low castes. The same consideration applies in a less strict way to fowls which roam about the village lanes; some individuals who eat the flesh of sheep and goats avoid eating fowls. Some of those who avoid eating the domestic pig have no objection to eating wild, jungle pork. Eating flesh is a mark of the lower castes because the taking of life in any form is a sin. The drinking of alcoholic beverages is again a mark of the lower castes. The Brahmans, who occupy the highest position in the caste hierarchy, avoid nonvegetarian food, including eggs, and also abstain from alcoholic drinks.

When a caste wants to rise in the hierarchy, it may adopt the Brahmanical diet. A striking example of this is provided by the Lingāyats. Some Smiths also have adopted the Brahmanical diet, but others have not. The latter are consequently regarded as inferior to the former.

There is a certain amount of surreptitious consumption of nonvegetarian food and alcoholic beverages, but this gets to be known eventually. For instance, one of my informants told me that the Traders in Rampura were vegetarians, but another cut in, 'So you think, but I once found a Trader woman throwing out domestic refuse in which there were bones'. Again, those who eat mutton would indignantly deny eating animals such as the domestic pig, the field rat, and water snake. It is alleged by some that the Peasants in neighbouring villages eat the domestic pig, but any public statement to this effect before a Peasant would lead to unpleasantness. It is also well known that the poorer Peasants in Rampura and around eat the field rat and water snake, but this again would be denied. Only a few of those who drink alcoholic beverages would admit to it.

Occupations also form a hierarchy. Butchery is a low occupation because the butcher kills animals for a living. In this area, only Muslims are butchers. It is true that all the nonvegetarian castes occasionally kill animals, but not for a living. Fishing also involves killing living creatures. Working with leather is a low occupation, because handling hide is defiling; such defilement may be related to the taking of life, and to the messiness of skinning and tanning. Only Untouchables work with leather. Herding swine is a low occupation because swine defile. The tapping and sale of toddy are low occupations because only low castes drink toddy; Western alcoholic drinks, which are consumed only by the wealthy, are not considered low. Although agriculture is an occupation common to all, Manu (X, 84) forbids it for Brahmans because the plough injures the earth and destroys living things. Brahmans in this area do not usually engage personally in agriculture, and even the richer Non-Brahmans have the actual work on the field done by servants.

The work of both the Barber and the Washerman involves handling dirt, and this makes the occupations of both unclean. Handling hair, and nails after they are separated from the body defiles the man who handles them. The Barber's touch consequently defiles a member of a higher caste. The Washerman handles soiled clothes, including menstrual clothes. It is interesting to note that both the Barber and the Washerman refuse to serve the Untouchables. On the other hand, the Brahman is extremely particular about purification after being shaved by the Barber. The spot where he and the Barber sat is washed with a solution of the purifying cowdung. Then a member of the Brahman's family pours several vesselfuls of water over him, wetting him thoroughly. The Brahman himself may not touch a bathing vessel before this. But some of the lower Non-Brahmanical castes are not very particular about taking a bath after being shaved by the Barber.

The clean clothes brought by the Washerman are purer than soiled clothes, but they are not pure enough for the Brahman to wear during worship. For this, either a silk cloth, or a cloth washed by a member of the Brahman's family, or a cloth washed by a Non-Brahman servant but subsequently dipped in water by a Brahman, is necessary.

Ideas of pollution do not attach themselves to working with iron, making pottery, basket-making, shepherding, or trade, excepting trade in low articles such as toddy and meat. There is no inherent reason why these occupations should be regarded as low. But the fact remains that they are, and the castes practicing them are unequal.

Castes in Rampura may claim higher rank not only by reference to the criteria of diet and occupation but also by reference to myths and to particular caste customs. Some identify themselves with positions in the order varṇa which sorts out castes into Brahmana, Kṣatriya, Vaisya, Sudra, and Untouchables. Unfortunately, however, the sociologist can take little comfort in these identification, for the hierarchical situation in any village or local area is quite unlike the varna view of the hierarchy. Nebulousness as to mutual position is one of the features of the caste system as it exists in fact, as distinct from the neat view which the traditional Brahmanical writers have put forth (see Srinivas, 1962).

Any attempt to arrange the castes of a village or local area into a hierarchy is therefore both difficult and fraught with risk. Any hierarchical list is necessarily tentative and arguable. But these considerations should not prevent an attempt, for the existence of a hierarchy and the preoccupation of village people with it are beyond doubt. The list in Table 2 represents an attempt to arrange the castes of Rampura in a hierarchy based on mutual ritual rank. Table 2 omits Muslims entirely, because their membership in another religion raises excessive uncertainties as to their hierarchical position.

Other ambiguities require that Smiths and Lingāyats be placed in separate columns rather than in one column with the other castes. Thus the Smiths of Rampura claim, on the one hand, that they will accept cooked food only from Brahmans. Smith men accept cooked food also from Lingāyats, but their women do not. On the other hand, most, if not all, of the other Hindu castes say that Smiths are inferior to them and, in support of this contention, point out that other castes do not accept cooked food and drinking water from Smiths. Even the Untouchables do not take food and water from Smiths. One reason for the Smiths' strange position is that they are said to belong to the Left-hand (*eḍagai*) division, while the bulk of the Non-Brahmanical castes, including the Holeya Untouchables, belong to the Right-hand (*balagai*) division. The Brahmans, and probably the Lingāyats as well, are in neither division. In those areas of peninsular India where Tamil, Kannada, and Telugu are spoken, the Non-Brahmanical castes are commonly grouped into Right- and Left-hand divisions, which were formerly bitter rivals (Thurston, 1909, III, pp. 117, 143; IV, p. 252). Castes belonging to the Left-hand division, such as the Smiths and the Mādiga Untouchables, were subjected to certain disabilities. For instance, Smiths in this area formerly could not perform their weddings within the village except in those villages where there was a temple to Kāli. The wedding procession of the Smiths was not allowed to pass through those areas where the high castes lived. No Smith was allowed to wear red slippers (*caḍāvu*). The marriage canopy of the Smiths was required to have one pillar less than the canopies of the others. The Smith is even today said to have 'one colour less' (*ondu baṇṇa kaḍime*) than the Right-hand castes, and there are myths which try to account for this saying.

TABLE 2
HIERARCHICAL LIST OF CASTES IN RAMPURA

Rank Group	Caste		
I	(A) *Hoysala* *Karnātaka* Brahman (B) *Mādhva*	*Lingāyat*	Smith
II	Peasant—Shepherd— Trader—Oilman—Potter— (A) Fisherman—Washerman—Barber— Basket-maker—Toddyman .... (B) Swineherd		
III	Untouchable		

Discrimination against the Smiths occurs everywhere in peninsular India, possibly as a result of their attempts in the past to rise high in the caste hierarchy by means of a thorough Sanskritization of their customs. Of the Kammālans (Smiths) of the Tamil country, Thurston writes (1909, III, p. 118):

The Kammālans call themselves Achāri and Paththar, which are equivalent to the Brahman titles Ācharya and Bhatta, and claim a knowledge of the Vēdas. Their own priests officiate at marriages, funerals, and on other ceremonial occasions. They wear the sacred thread. . . . Most of them claim to be vegetarians. Non-Brāhmans do not treat them as Brāhmans, and do not salute them with namaskāram (obeisance).

The Madras Census of 1871 notes that the Kammālans '"have always maintained a struggle for a higher place in the social scale than that allowed to them by Brahmanical authority . . . . There is no doubt as to the fact that the members of this great caste dispute the supremacy of the Brahmins, and that they hold themselves to be equal in rank with them". John Fryer, who visited India in 1670, seems to refer to this attitude' (cited in Ghurye 1932, p. 6). The Smiths' attempt to rise to the top of the hierarchy in the Tamil country by Sanskritizing their customs seems, as in Mysore, to have earned them only the combined hostility of most of the other castes.

The Lingayats are another Non-Brahmanical caste of Rampura who question the supremacy of the Brahmans. They worship the deity Śiva in his several manifestations, are strict vegetarians, and abstain from alcoholic beverages. They have their own priests and do not call in the Brahman priest. Some of them refuse to eat food cooked by Brahmans. Most Non-Brahman castes eat food cooked by the Lingāyats. The Brahmans do not, however, accept cooked food or water at the hands of Lingāyats.

The other castes in the village may be put approximately into a hierarchy as is shown in Table 2, but I am convinced that such a ladder-like arrangement is not a perfect way of representing the situation. For instance, in Table 2 the Holeya Untouchable is shown as occupying the bottom of the hierarchy. But he would claim that he was not inferior to the Smith and to the Mārka Brahman. In support of his claim he would point out that he belongs to the Right-hand division, while the Smith belongs to the Left-hand division, and that he does not accept cooked food and water either from the Smith or from the Mārka Brahmans of neighbouring villages (cf. Thurston 1909, I, pp. 367–8).

Frequently the position claimed by a caste differs from the position conceded to it by others, and the sociologist has either to accept one of these claims or to construct his own picture of the hierarchy. The sociologist's construction cannot claim complete objectivity, for it involves the evaluation of statements made by his informants. But it is less subjective than the claims of any one of the castes themselves. The sociologist, for instance, points out that, while the Holeya Untouchable claims to be superior to the Smith, the former has certain disabilities which the Smith does not have. The Untouchable has to live apart from the other castes, and he may not bathe or take water in a river at a point higher than that utilized by a member of the other castes. Similarly, he may not take water from or bathe in the tank but must use the tiny canals on the other side of the road which take water away from the tank into the fields. The Untouchable may not come into the temples of the higher castes, while the Smith may. And so on. But it must be noted here that the Smith too has, or had, certain other disabilities which have already been mentioned. The sociologist evaluates these two kinds of disabilities and says that one kind puts a caste into a lower position than the other.

The castes of the middle group II(A) span a considerable structural distance without definite lines between any two of them. Some pollution is involved in the work of Washermen and Barbers, and also in the work of those Toddymen who handle toddy or the leaves of the toddy palm. Washermen, Barbers, and Toddymen are therefore placed near the bottom of the II(A) group of castes. No such pollution is involved in the work of any of the other castes of this subdivision.

Peasants and Shepherds in Rampura regard themselves as standing higher than every other caste except Brahmans and Liṅgāyats, and they are accordingly placed at the top of the middle group of castes. Here we come across an important principle of caste hierarchy that is not sufficiently, if at all, acknowledged—the presence of local factors which influence the structure of the hierarchy. In Rampura the Peasants are the dominant landed caste, and the Shepherds are only next to them in strength and importance. This local dominance gives the Peasants a high status among the castes in the middle division. The local numerical strength of a caste and the amount of land it owns are not the only factors, however. The actual occupation locally pursued and the extent of Sanskritization are also important. Thus there is one Toddyman of Rampura who occupies a higher position by virtue of having given up the direct handling of toddy and having taken to new occupations. This underlines the fact that the hierarchy is everywhere influenced by local factors: since local factors may change over a period of time, the hierarchy is also dynamic.

Definitely beneath these castes, but above the Untouchables, are the Swineherds, who are consequently put into a lower sub-divison, II(B). The Swineherd's herd swine, eat pork, and drink toddy. Their touch defiles. The Peasant headman once refused to cut a mango fruit with a Swineherd's knife because he feared that the knife might have been used to cut slices of pork. No one eats food cooked by the Swineherds, or drinks water from a vessel touched by them.

There is one final and complex point. The hierarchy which is presented in Table 2 has ritual considerations as its basis. That is, the castes are arranged in a particular order on the basis of ideas regarding pollution. But there is, at least nowdays, a certain discrepancy between the hierarchy as it is conceptualized by the people and as it exists in behaviour. Discrepancy is due to the fact that, in conceptualizing the hierarchy, ritual considerations are dominant, while in the day-to-day relationships between castes economic, political and 'Western' factors also paly an important part. Thus the relation between the poor Brahman priest and the rich Peasant headman of the village is a complicated one, the Brahman being aware of the secular power of the headman, and the headman showing deference to the Brahman's ritual position. The local Untouchable servant or tenant is treated as an inferior by a Peasant, but, when the same Peasant meets an Untouchable official, he shows respect, although grudgingly. Thus there are ritual, economic, political, and 'Western' axes of power, and any single point of contact between individuals belonging to different castes is governed by all these axes which are present in the point. All the axes may be said to be implicit in any single act of contact.

Patrons and Clients

No account of a village social system in this part of Mysore State can be complete without reference to certain institutionalized vertical relationships between individuals and, through them, between families. These relationships include the relationships of master and servant, landowner and tenant, and creditor and debtor; they may be viewed collectively as the relationships of patrons and clients. Some of these relationships link persons of different castes, and others may link persons who are rich with persons who are poor, but all these relationships are essentially unequal.

One of these vertical relationships is now defunct, although it was an important part of Rampura's social system before World War I. This is the relation of traditional servantship which prevailed between Untouchables and Peasants. The traditional Untouchable servant was called the 'old son' (*halimaga*) of the particular Peasant whom he served. This traditional servant had certain well-defined duties and rights in relation to the master and his family. For instance, when a wedding occurred in the master's family the men of the servant family were required to repair and whitewash the wedding house, put up the marriage canopy before it, chop wood to be used as fuel for cooking the wedding feasts, and do odd jobs. The servant was also required to present a pair of leather sandals (*cammāḷige*) to the bridegroom. Women of the servant's family were required to clean the grain, grind it into flour in the rotary quern, grind chilies and turmeric, and do several other jobs. In return for these services, the master made presents of money and of cooked food to the servant family. When an ox or a buffalo died in the master's household, the servant took it home, skinned it, and ate the meat. He was required, however, to make out of the hide a pair of sandals and a length of plaited rope for presentation to the master.

Many Untouchable families and Peasant families were bound together in enduring ways by the institution of traditional servant-ship, despite the wide separation of the two castes in the hierarchy. Since the Untouchables are and were very poor, it is likely that some of the traditional servant-master relationships were reinforced by tenancy, contractual servantship, debtorship, and other ties as well.

Jita servantship may be termed 'contractual' servantship, to mark it off from traditional servantship. Under it a poor man contracts to serve a wealthier man for one to three years. The terms of the service, including the wages to be paid by the master, are usually reduced to writing. The master advances, at the beginning of the service, a certain sum of money to the servant or his guardian, and this is worked off by the servant. Usually no interest is charged on the advance unless the servant tries to run away or otherwise break the contract. The sum paid is exclusive of food and clothing, which it is the master's duty to provide. Frequently, before the period of the service runs out, the servant or his guardian borrows another sum of money and thus prolongs the service. Formerly it was not unknown for a man to spend all his working life between ten and seventy years of age in the service of one master. In one case a servant lived with his joint family, numbering over a dozen, in the house of his master, who was also an agnatic kinsman. On the death of the servant the corpse was accorded the honour of a burial in the master's land, near the graves of the master's ancestors.

Some members of most castes in Rampura are involved in con-tractual servantship, either as servants or as masters. In 1948 there were fifty-eight servants in Rampura. These servants came from every caste except the Brahman and Liṅgāyats castes, and included fourteen Untouchables. Masters were found in every caste excepting in the Untouchable caste, which ranks at the bottom, and in the Smith castes, whose members are assisted at their work by relatives and customers. Hindus and Muslims are bound together by contractual servantship, for Muslim masters invariably employed Hindus as servants, while Muslim servants served only Hindu masters.

The bond between master and servant is intimate. Contractual servantship is often only one of the bonds prevailing between the two families. Sometimes a master employs a man as tenant on condition that he agrees to having his son or younger brother work as servant in the master's house. Caste, kin, and other ties frequently strengthen the tie between master and servant. When a servant works for a master long enough, he tends to be treated as a member of the family. It is not unknown for even an Untouchable servant to fondle his Peasant master's child, in spite of the theoretical ban against such contact. In fact, the conditions of service frequently require the violation of rules regarding pollution.

The master is, in certain circumstances, regarded as responsible for the acts and omissions of the servant, though there is no clear and explicit formulation of the doctrine of vicarious responsibility. An Untouchable servant of the headman was once accused of being abusive to a Peasant. The servant said in defence of his conduct that the Peasant had been diverting water which ought to have gone to the headman's field. When the headman's second son was called to arbitrate the case, it was clear that he secretly approved of what his servant had done. But he had to appear impartial, and the wrong of which his servant had been accused was a serious one. Had the Untouchable been acting in his own right, it is likely that he would have been belaboured by the Peasant. But, as things stood, the Peasant had to rest content with simply lodging a complaint.

A rich man does not personally cultivate but has his young sons or servants or tenants manage the agricultural work. In the few top families in the village, even the young sons do not personally handle the plough, though they regularly go to the fields to supervise the work of servants. Servants are cheaper than tenants, but they require close and regular supervision; tenants require no supervision and possess their own ploughs and oxen. Landowners who are resident in Rampura exact a day's *corvée* from their tenants during the transplantation season. A landowner may also demand his tenants' labour and support on other occasions.

The landonwer-tenant relationship occasionally cuts across caste barriers, and this is more common when the landowners are permanently absent from the village. The relationship between landowner and tenant is also an intimate one. Like all intimate relationships, it is frequently marked by conflict. Tenants are heard complaining against the exploitation of the resident landowners; they have begun to feel that absentee landowners have no right to receive income from the land. There is an acute shortage of riceland in this area, and where landowner and tenant, or competing tenants, belong to different castes, the struggle over land may be seen as a clash of castes.

Seasonal fluctuations in the demand for labour in this rice-growing area contribute to the forging of other interpersonal ties which may ignore the barriers of caste. A man finds it difficult to obtain labour when he wants to, especially during transplantation and harvest. Time and numbers are both crucial factors on these occasions. Then the village puts into the field all its available labour force, including men and women. During the harvest, men and women come also from a few neighbouring villages to cut the stalks, thresh the grain, rick the straw, and cart the grain away for storage.

Servants, and even tenants, help a man in coping with the work of transplantation and harvest, but they are not enough. Extra labour has to be employed. This may be either paid for in cash or secured on the basis of a reciprocal arrangement with other cultivating families. But securing labour on the basis of reciprocity depends on the ties of kinship, caste neighbourliness, and friendship. A man must be friendly and ready to help another with his labour, time, resources, and money, if he wants others to help him.

A word that is constantly heard in the village is *dākṣiṇya*, which may be translated as 'obligation'. Because of 'obligation', one is frequently called upon to do things one does not want to. Every relationship between two human beings or groups is productive of 'obligation', and gives each of them a claim, however vague, on the other. If A once refuses to do what B wants him to do, then B may sometime refuse to do what A wants him to do. A poor man can put others under his obligation only by giving his personal labour and skill. But a rich man has many devices: he can oblige others by lending them money, by letting them land, by speaking to an official or big man on their behalf, or by performing acts of generosity. Thus a rich man is able to put many persons under his obligation. Every rich man tries to 'invest in people', so that he can on occasion turn his following to political or economic advantage.

The several relationships between a master and his *jīta* servants, a landowner and his tenants, a creditor and his debtors, and finally between a rich man and his dependents, may all be subsumed under a single relationship: patron and client. I use the term 'patron' in its loose, dictionary sense to mean 'one who countenances, or gives influential support to person, cause, art, etc.' Such a subsumption is legitimate, as, usually, it is a rich landowner who employs *jīta* servants, lets some of his land to tenants, lends money, and otherwise helps people. Every important man gathers around him a number of people, who may be his relatives, caste-folk, tenants, servants, debtors, or potential debtors, those who vaguely hope to receive some advantage from him, and those who just enjoy basking in the warmth of a patron's power. The following of a patron crosses to some extent the barriers of caste. The relation between patrons is frequently one of rivalry, and such rivalry is expressed on various oaccasions, ritual as well as secular.

During the summer of 1952 I tried to sort out the following of each major patron in Rampura. This was a delicate task and had to be conducted with a good deal of caution. Any open inquiry into the following of each patron was bound to be interpreted as an attempt to expose the seamy side of village life. Not only would such an attempt have been resented, but the majority of people would have refused to label themselves as clients of any one patron, fearing that this would make enemies of other powerful people. I was forced to rely on my own knowledge, supplemented by questioning a few trusted informants. The result is not wholly satisfactory, but I am presenting it here for what it is worth.

My list accounts only for a part of the population of Rampura. I was told that the Untouchables were all clients of the village headman, who may be called Patron I. While they all do follow him in a general way, I am aware that a few Untouchables have also special relationships of dependence upon other patrons. Such other relationships are bound to affect adversely their clientship under Patron I. Multiple relationships of dependence also create one of the chief difficulties in ascertaining clientship. Only the hard core of a patron's following is willing clearly to declare its allegiance to one patron, while many clients have a marginal affiliation to more than one patron. Marginal clients give fluidity to the followings of the various patrons as they shift their allegiance from one to another over a period of time.

In the list which follows, only the most important patrons are mentioned. Minor patrons who are themselves the clients of greater patrons are ignored.

In addition to his Untouchable clients, the headman (Patron I) has a following of fifty-eight families. These comprise twenty-one Peasants, eight Shepherds, ten Muslims, five Potters, three each from Trader, Smith, and Lingāyat castes, two Oilmen, and one each from Brahman, Washerman, and Toddyman castes. Patron II has a following of nineteen Peasants, five Shepherds, three Oilmen, two Muslims, and one Smith. Patron III, who is a junior member of the same lineage as Patron II, has sixteen Peasants, one Brahman, and one Muslim as clients. Patrons IV and V have followings of six and four Peasants, respectively.

There is a wide gulf between Patron I and the other patrons, a gulf which has increased since 1949 as a result of a split in the biggest Peasant lineage in Rampura. One part of this lineage is led by Patron II and the other by Patron III. Patron I, as the official headman of the village, has some influence with government officials and Congress leaders. He is far wealthier than the other patrons, and his joint family has the tradition of leadership of the village since its founding. As a result of his dominance, Rampura shows a measure of unity and harmony which does not prevail in neighbouring villages.

The word 'party' has become a Kannada word. Every administrator and politician speaks of 'party politics' in villages, and even villagers are often heard saying, 'There is too much "party" in such and such a village'. The coming of elections has given fresh opportunities for the crystallization of parties around patrons. Each patron may be said to have a 'vote bank' which he can place at the disposal of a provincial or national party for a consideration which is nonetheless real because it is not mentioned. The secret ballot helps to preserve the marginal affiliation of the marginal clients.

Structural Unity of the Villages

Rampura is a well-defined structural entity which commands the loyalty of all who live there, irrespective of their affiliation to different castes.

There are many bonds opposing the divisiveness of caste in Rampura. One is physical: like other villages in the plains of eastern Mysore, Rampura is a close cluster of huts surrounded by fields. Each such village is cut off from other villages and from towns owing to the lack of roads. The degree of isolation was even greater in the past, when government was mainly a tax-collecting body.

Each village is a tight little community in which everyone is known to everyone else and in which a great deal of experience is common to all. Agricultural activities in which the vast majority of the villagers are engaged impose the same activity upon all of them at any given period in the year. Hindu festivals are common to the bulk of the inhabitants. A drought or excess of rain is of common concern to all. Formerly, during an epidemic of plague, or cholera or smalllpox, the village was evacuated, and temporary huts were put up at some distance. Everyone returned to the village only after the epidemic had died out.

Patriotism for one's village is common. Patriotism finds expression positively in the enumeration of Rampura's virtues, and negatively in the criticism of neighbouring villages. It also manifests itself occasionally in opposition to the government. During the summer of 1948 the agricultural department passed an order stating that fishing rights in village tanks would be auctioned thenceforth. This produced a protest at once from everyone, including the headman, and a petition was immediately drawn up and dispatched to the government. The villages felt that the government was encroaching on their rights to fish in the village tank when they wanted to. An auction was held a few days later, but no one bid. Care had been taken also to send word to neighbouring villages not to bid. Thus a silent act of non-co-operation nullified a government order.

The unity and solidarity of the village emerge most clearly in relation to the government. A criminal from the village is afforded protection as long as he operates outside the village, and as long as it is not too risky to hide him. There are occasionally fights between villages, but these are limited by the fact that individuals and families in the quarrelling villages have numerous contacts with each other. A fight causes hardship to many.

The unity of the village finds further expression in ritual contexts. The entrance to a village (*rādu bāgilu*), usually unmarked, receives ritual attention on certain occasions. Every village has a temple to the goddess Māri, who presides over epidemics, and she is propitiated in order to drive an epidemic out of the village. It is believed that if the corpse of a man or woman suffering from leucoderma is buried in the ground, a long drought will result. Such corpses are either floated down a river or exposed in stone structures (*kallu sēve*) on hilltops. It is believed that the misconduct of a priest may result in the deity's leaving the local temple and settling down in some other village.

Every village has a hereditary headman, an accountant, servants belonging to the Untouchable castes, and watchmen (Kāvulu). These functionaries act for the whole village and not for any one section of it.

The village may, then, be described as a vertical entity made up of several horizontal layers each of which is a caste. Yet I believe that the physical imagery involved in this description may be a handicap in thinking about intercaste relations. For testing the vertical unity of the village a crucial question is, 'How far does the unity of the village really include polar groups like the Brahmans and Untouchables, and a peripheral group like the Muslims?' Much to my regret the importance of this question did not occur to me until I had started to write up my field data.

In October, 1947, a fight occurred between Kere and Bihalli at Gudi, at the annual festival of Mādeśvara. I have an account of the fight, obtained about six months later, but it never occurred to me to ask to which castes the participants belonged. I know that the bulk of them were Peasants and members of other castes of the middle range. But I do not know if Brahmans, Liṅgāyat, Untouchables, and Muslims were also involved in it. The question which is important to ask is, 'Would a Brahman, Untouchable, or Muslim from either village be attacked merely by virtue of his belonging to it?' My own guess is that a Brahman would not be attacked, because of his position in the hierarchy. An Untouchable would be involved more because of his position as client to a high-caste patron than by virtue of his membership in the village. A Muslim would be in a similar position.

In the Non-Brahmanical village festivals, the Brahmans, Untouchables, and Muslims play at best an unimportant part. The co-operation of the Untouchables and Muslims is, however, sought in the work of the festival, and the Brahman is paid rice, lentils, salt, chilies, tamarinds, and vegetables—the ingredients of a meal.

Summary and Conclusion

I hope that I have given some idea of the nature of the ties that run across the lines of caste in a multicaste village. While the divisive features of caste have previously received notice, the links that bind together the members of different castes who inhabit a village, or a small local area, have not been adequately emphasized. Many features of village life tend to insulate castes from each other: endogamy, the ban on commensality, the existence of occupational specialization, distinctive cultural traditions, separate caste courts, and the concepts of pollution, *karma* and *dharma*. But there are counteracting tendencies too.

Occupational specialization requires interdependence among the castes, a fact which is dramatized in the annual grain payments made to the serving castes. Yet the availability of agricultural occupations as alternatives for members of all castes at the same time serve to underwrite occupational specialization. Along with migration and production for sale in weekly markets, the alternative of agriculture offers a means for absorbing excess persons from the nonagricultural castes. It makes possible the opening of new land by any caste group, and during times of increasing population it prevents widespread confusion by keeping the surplus population alive. Finally, acquisition of land, along with Sanskritization, makes mobility in the system possible.

Occupational specialization is important in other ways too. It gives each group a vested interest in the system as a whole, because under it each group enjoys security in its monopoly. Monopolies are jealously safeguarded by various means. But the families enjoying a monopoly are also competitors, which means that kinship tensions and economic rivalries may drive each family to seek friends outside the caste.

There are vertical institutions which bring together families and individuals belonging to different castes. Such institutions are *jita* service, tenantship, debtorship, and clientship. As land was the principal form of wealth in the traditional economy, all these institutions eventually depended upon the private ownership of arable land.

Local methods for settling disputes reveal the part played by the elders of the dominant caste. These elders, standing in an intermediate caste position, wield economic and political power over all the minor castes. These elders are the guardians of the social and ethical code of the entire village society. They represent the vertical unity of the village against the separatism of caste.

In sum, the village is a community which commands loyalty from all who live in it, irrespective of caste affiliation. Some are first-class members of the village community, and others are second-class members, but all are members.

REFERENCES

CENSUS OF INDIA, 1942, *Census of India, 1941*. Vol. XXIII: *Mysore, Part II—Tables* (Government Press, Bangalore).

GHURYE, G. S., 1932, *Caste and Race in India*, Kegan Paul & Co., London.

GOUGH, E. KATHLEEN, 1952, 'The Social Structure of a Tanjore Village', *Economic Weekly* 4: 531–6. (Bombay).

MARRIOTT, MCKIM, 1952, 'Social Structure and Change in a U.P. Village', *Economic Weekly* 4: 869–74. (Bombay).

SRINIVAS, M. N., 1952, *Religion and Society among the Coorgs of South India* (Clarendon Press, Oxford).

——, n.d. 'Varna and Caste', in M. N. Srinivas, *Caste in Modern India and Other Essays* (Asia Publishing House, Bombay).

THURSTON, EDGAR, 1909, *Castes and Tribes of Southern India*, 7 vols. (Madras Government Press, Madras).

The Dominant Caste in Rampura[1]

I

The concept of the dominant caste is crucial to the understanding of rural social life in most parts of India. Whether analysis is to be made of the hierarchy of a multi-caste village, the settlement of a dispute at the level of village or caste, or the pattern of Sanskritization among the several castes of an area, a study of the locally dominant caste and the kind of dominance it enjoys is essential. Occasionally a caste is dominant in a group of neighbouring villages if not over a district or two, and in such cases, local dominance is linked with regional dominance. Such linkage also exists when the caste which is locally dominant is different from the caste which is regionally dominant.

I stumbled on the importance of the idea of dominant caste only in 1953, after I had made two field trips to Rampura, a multi-caste village about 22 miles southeast of Mysore City in South India, and the present analysis is based on material which was collected previously. A full understanding of the dominance which a caste such as the Peasants (Okkaligas) enjoy needs a study of the entire region over which they are dominant, and over a period of time. I regret that I do not have the data for such an analysis. My analysis would have been even sketchier but for the fact that in 1952 the headman of the Peasants in the neighbouring village of Kere loaned me several documents which related to the settlement of disputes in the Kere area over a period of forty years. These documents referred to villages in Kere *hobli* (an administrative division referring to a group of 20–50 villages) which is different from the *hobli* to which Rampura belongs. But as Peasants are dominant in both the areas, and as culturally the two areas are quite close to each other, I have made use of the Kere documents in order to clarify the concept of the dominant caste.

[1] This paper was read before a Departament of Anthropology Seminar at the University of Chicago in the last week of May, 1957. I thank the Rockefeller Foundation for a generous fellowship which enabled me to devote the greater part of the academic year 1956–7 to the analysis and writing up of my Rampura material. A full acknowledgement will be made when my book on Rampura is published.

I have elsewhere defined a dominant caste in the following words:

A caste may be said to be 'dominant' when it preponderates numerically over the other castes, and when it also wields preponderant economic and political power. A large and powerful caste group can be more easily dominant if its position in the local caste hierarchy is not too low [Srinivas, 1955, p. 18].

However, the above definition omits an element of dominance which is becoming increasingly important in rural India, namely, the number of educated persons in a caste and the occupations they pursue, I have called this criterion 'Western' (Srinivas, 1955, p. 26), since Western and non-traditional education is the means by which such dominance is acquired. Villagers are aware of the importance of this criterion. They would like their young men to be educatead and to be officers in the Government. As officers they are expected to help their kinsfolk, castefolk and co-villagers.

When a caste enjoys all the elements of dominance, it may be said to be dominant in a decisive way. But decisive dominance is not common; more frequently the different elements of dominance are distributed among the castes in a village. Thus a caste which is ritually high may be poor and lacking strength in numbers, while a populous caste may be poor and ritually low.

The Peasants in Rampura enjoy more than one element of dominance. Numerically they are the biggest caste with a membership of 735, while the next biggest is the Shepherd with 235, followed by the Muslim, 179, and the Untouchable, 125. The biggest landowners are among the Peasants, and the Peasants together own more land than all the other castes put together. There are also more literates and educated men among Peasants than among the others. In 1948 there were three Peasant graduates and a single Lingāyat lawyer employed by the Government. The three most important patrons in the village were also Peasants.[2] All of them owned land and loaned money. The official Headman of the village was one of these; he was the biggest landowner, owned two buses, and had built a few rental houses in a nearby town. The second was Nadu Gowda,[3] who had kept two shops and a small rice mill. The third was Nadu Gowda's agnatic cousin Millayya, who owned a big rice mill.

[2] For an elaboration of the concept of patron, see my 1955 essay.
[3] Here Nadu Gowda is the name of a Peasant; it is usually the name of the hereditary headman of the Peasant caste in a *hobli*-capital.

The ritual rank of Peasants is not very high. While they do rank above the Untouchables and such low castes as the Swineherd, they are well below Brahmins and Lingayats. In terms of varna they are Shudras, the fourth category in the all-India hierarchy. But this does not mean much in Rampura, as there are no 'genuine' Kshatriyas or Vaishyas. (The local trading caste of Banajigas are not accorded the status of the 'twice-born' Vaishya.)

While it is true that Peasants are not ritually high, they command respect from everyone in the village including the priestly castes of Brahmins and Lingayats. The members of the latter castes consult one or another of the Peasant patrons on important occasions. Even on ceremonial occasions, outside pollution contexts, Peasants are shown respect by Lingayats and Brahmins. Everyone is aware of the dominant position which Peasants occupy in Rampura.

Over the last fifty years or more, the dominance of Peasants has increased in Rampura. The available evidence indicates that in the early years of this century Brahmins owned a considerable quantity of irrigated land in the village. The Brahmins were the first to sense the new economic opportunities opened to them through Western education, and they gradually moved to the towns to enter the new white-collar professions. Urban living, the cost of educating children, and the high dowries which the new education and economic opportunities had brought about, gradually caused the Brahmins to part with their land. Much of this land passed to non-Brahmins, especially the Peasants, during the years 1900–48.

In the different parts of South India shortly after World War I there began what may be called the Non-Brahmin Movement. At the end of World War I, most of the important posts in the Government of Mysore were held by Brahmins, and non-Brahmin leaders realized that they must get Western education if they wanted position and power. Agitation was started for the institution of scholarships to help non-Brahmin youths study in schools and colleges, for reservation of seats for non-Brahmins in medical and technological colleges, and for preference in appointments to government posts. The non-Brahmin agitation succeeded, and gradually a number of rules discriminating against the Brahmins were evolved by the Government of Mysore. As a result of these measures there has come into existence since the late thirties a Western-educated non-Brahmin intelligentsia (see Srinivas, 1957).

This Non-Brahmin Movement is relevant to the understanding of the situation in Rampura. It was in the thirties that the leaders among Peasants in Rampura and the neighbouring villages began to think of higher education for their sons. Contact between the Peasants in Rampura and Peasant politicians and officials outside increased in the forties; furthermore, contact with the towns increased generally, and a few Peasants and Lingayats frequently went to Mysore and Bangalore to secure permits and to buy machinery and other goods.

The Brahmins and Lingayats in Rampura provide an instance of ritual dominance existing by itself, unaccompanied by the other forms of dominance. Neither caste is numerically strong nor is it wealthy. But some families in these two castes, namely, the Brahmin priest of the Rama temple and the Lingayat priests of the Madeshwara and Basava temples, are quite well off by village standards. The main source of income for these families is from the land with which the temples have been endowed, while a subsidiary but not unimportant source is the gifts in cash or kind which the devotees make to the priests whenever they visit the temples or during harvest. The eldest son of the Rama priest is employed in the Integral Coach Factory in Perambur (Madras) while, as mentioned earlier, one of the Lingayat priests practices as a lawyer in a neighbouring town.

But when a caste enjoys one form of dominance, it is frequently able to acquire the other forms as well in course of time. Thus a caste which is numerically strong and wealthy will be able to move up in the ritual hierarchy if it Sanskritizes its ritual and way of life, and also loudly and persistently proclaims itself to be what it wants to be. It is hardly necessary to add that the more forms of dominance which a caste enjoys, the easier it is for it to acquire the rest.

What I have said above applies only to caste Hindus; untouchability constitutes a serious obstacle to group mobility. Untouchables in Rampura are either landless labourers, tenants, or very small landowners. They started going to school only in the thirties. In 1948, Untouchable leaders from outside were going around asking Untouchables in the Rampura area to try to shake off the symbols of untouchability. In the neighbouring village of Bihalli, for instance, Untouchables decided to give up performing services such as removing the carcasses of dead cattle from the houses of the higher castes, beating the tom-tom at the festivals of village deities, and removing the leaves on which the high castes had dined during festivals and weddings. The Bihalli Peasants became annoyed at this and beat up the Untouchables and set fire to their huts. A similar attempt by the

Kere Untouchables was nipped in the bud by the local Peasants.

The dominant caste of Peasants in Rampura is plainly opposed to the emancipation of Untouchables. Government efforts to improve the position of Untouchables are often frustrated by the leaders of the locally dominant caste. Thus, in 1948, the Government of Mysore sanctioned a sum of money to enable Untouchables in Rampura to have tiled roofs instead of thatch. The grant was administered through the Headman. The Untouchables later complained that the Headman did not readily give the money, and then only a small part of what he should have given. The Peasants, on the other hand, said that the Untouchables had spent the money given to them on toddy, and that this showed that Untouchables could not be improved.

Thus, while the Government of India and Mysore want to abolish untouchability, and the Untouchables themselves want to improve their position, the locally dominant caste stands in the way; its members want the Untouchables to supply them with cheap labour and perform degrading tasks. They also resent the idea that Untouchables should use their wells and tanks, and worship in their temples. They have the twin sanctions of physical force and boycott at their disposal. It is true that the Untouchables can enforce their rights with the aid of the Police and Law Courts, but there are many considerations which come in the way of taking such a drastic step.

II

The numerical strength of a caste influences the kind of relations which it has with the other castes, and this is one of the reasons why each multi-caste village to some extent constitutes a unique hierarchy. No two villages are identical either in the number of castes represented or in the numerical strength and the wealth of each resident caste. In fact, the same caste may occupy different positions in neighbouring villages. For instance, in Kere, Fishermen are not allowed to take their wedding and other processions into the streets in which Brahmins and Peasants live, whereas in those villages in Malavalli Taluk where Fishermen are in the majority, no such disabilities affect them. There are other instances where the position of a caste is influenced by considerations such as the amount of land owned by its members and the degree to which its way of life is Sanskritized and Westernized. When the same caste occupies different positions in different villages, the segment of the caste which is occupying the lower position will be stimulated to move up in the local hierarchy. Members of the minority castes in Rampura occasionally told me with pride that in a particular village their castefolk were numerous and wealthy. They were trying to identify themselves with people whom they regarded as having a higher position than themselves.

Where a caste is numerically strong, its members have the assurance that the other castes in the village will not be able to subject them to any insult or exploitation. (The Untouchables are to some extent an exception to this.) The capacity to 'field' a number of able-bodied men for a fight and a reputation for aggressiveness are relevant factors in determining the position of a caste *vis-à-vis* the other castes. Considerations of power do prevail; the system adjusts to the situation that obtains in any single village.

I visited Kere a few times during the summer of 1952 and found that the Brahmins there were suffering from a sense of insecurity. In the General Elections held a few months previous to my visit, the members of the family of the Brahmin accountant had actively canvassed for a candidate who was not residing in Kere. This enraged the local candidate, a very powerful Peasant. The outsider won and the defeated candidate freely expressed his dislike of Brahmins in general and the accountant's family in particular. He even said that he wanted the Brahmins to leave Kere. A Brahmin doctor in Kere told me that where they are few in number, Brahmins had no future in the village. He thought that Brahmins ought to migrate to villages and towns in which they were represented in some strength.

Two incidents which occurred in Rampura in the summer of 1952 further drove home to me the sense of insecurity prevalent among members of minority castes. One was a dispute between a Potter and a Lingayat Priest in which the latter told an influential Peasant friend that the Lingayats were in a minority in Rampura and that it was up to the Peasants to see that they were not humiliated. The implication was that had there been enough Lingayats, they need not have depended upon the Peasants to secure them justice.

During the summer of 1952, a Rampura Shepherd sold all his land in the village and his share of the ancestral house in the village. It was stated in the sale deed that he was leaving Rampura because only a few Shepherds (actually 175) were living there, unlike his affinal village in which they were preponderant. His action was unpopular, and it was widely rumoured that the real reasons for his leaving the village were his inability to get along with his brothers and his friendliness with his wife's kin. But it is significant that those who drafted the document regarded moving into a village where his caste was represented in strength as a proper and sufficient reason for selling his land and house in his natal village.

⸝ Statements are often made by members of minority castes that they have no protection against bullying and exploitation on the part of men of the dominant caste. The members of the non-dominant castes may be abused, beaten, grossly underpaid for work done, or their women required to gratify the sexual desires of the powerful men in the dominant caste.

It is not unlikely that the concentration of the members of a caste into a ward (a feature of village life all over India) adds to their sense of security. While this practice is related to ideas of pollution and purity, this is not the whole story. A man feels safer in the *keri* (ward) of his caste.

A patron's following can be made to yield him economic and other benefits. Patrons from the dominant caste can secure a larger number of followers than patrons from non-dominant castes. The rural patrons are 'vote banks' for the politicians, and during elections they are approached for votes. In return, patrons expect favours—licences for buses and rice mills, and seats in medical and technological colleges for their kinsfolk. The existence of such links between patrons and politicians establishes a continuum between rural and urban forces, making each responsive to the other.

III

Disputes are referred to patrons for settlement, and where there is a decisively dominant caste in a village, the biggest patrons usually come from that caste.

The word panchayat has been used to include all judicial and executive boides in rural areas. I shall here restrict the use of the term to the official panchayat only, i.e. the assembly of village elders constituted according to an Act of the Provincial or State Government. It is the official panchayat which is entitled to levy a tax on every house, house-site, and shop in the village. The money which the official panchayat collects is to be used to provide the village with drains and street lights, to improve the village well, tank, or temple, and for other similar purposes. Generally, the State Government lays down the procedure by which the official panchayat should be constituted. There was an elected official panchayat in Rampura in 1948 constituted under the Mysore Government's Village Panchayat Act of 1926, under which every village was required to have a panchayat with a minimum of twelve members, not less than half being elected. The Chairman was usually nominated, but the Deputy Commissioner, the official in charge of a District, was given the discretionary power to allow a village to elect the chairman annually. He did this only when he thought a particular panchayat deserved the honour. In 1947, Rampura was given that honour, and the immediate result was a keen contest for the chairmanship. The Headman was nearly unseated in the struggle. The Headman became Chairman again in 1948, but this was because no election was held—even though the official report stated that an election had been held at which the Headman was elected Chairman.

In 1948 the official panchayat included leaders from every numerically significant caste, including Muslims and Untouchables. It also included the Brahmin village accountant. The panchayat minute book reported the holding of a meeting once a month, but this was an exaggeration. The members even reported one absentee at every meeting to give verisimilitude to their minutes. During my entire stay in Rampura (in all, about 13 months), the official panchayat met only once. The Chairman of the panchayat either made all the decisions by himself or in consultation with his great friend, fellow-casteman, and relative, Nadu Gowda.

The official panchayat is usually dormant and becomes active only on certain occasions. It is the traditional and unofficial panchayat—here called council—which is active in the settlement of disputes. The membership of the council varies from village to village and from context to context. It may on occasion include all the leaders of the numerically significant castes, or it may include only the disputants concerned and a patron like the Headman or Nadu Gowda. A patron usually acts on a complaint received from someone, except when he himself is the aggrieved party. He may feel that it is not necessary to call anyone else or he may ask the disputants to request a few other patrons to come together. Thus when a Peasant brought a complaint before Nadu Gowda against a Smith woman, saying that her dog had eaten his lamb, he did not feel it necessary to invite anyone else to 'sit on the bench' with him. Village councils are informal and flexible, and there is no hard and fast rule about who should sit on them. A great deal is left to the discretion of the patrons.

Traditional councils may be divided into caste councils and village councils, depending on the kind of issue before the patrons. This distinction is not absolute; there is occasional overlapping of jurisdiction. Village councils have jurisdiction in matters such as: 'Who stole grass from X's field?', 'Who set fire to Y's straw-rick?' and 'Is Z speaking the truth when he says that P owes him Rs 100 and not Rs 50?' Caste councils decide such questions as: 'Should R be thrown out of caste for having sex relations with an Untouchable woman?' and, 'Should J be granted a divorce from M?'

A caste council usually has jurisdiction over disputes among members of a single caste. In a dispute in which members of different castes are involved, patrons from the concerned castes and a few patrons from the dominant caste form the council. The patrons of the dominant caste have jurisdiction over all the castes living in the area. Such jurisdiction is invoked through the pre-existing bonds of patron-and-client, kinship, or friendship.

It is necessary here to comment briefly on the role of the Headman and Nadu Gowda in the settlement of disputes in Rampura. Both are members of the dominant caste, heads of large lineages, landowners, money-lenders, and patrons. The Headman is also the holder of an hereditary government post which gives him power and influence in the village. His joint family has considerable prestige in the area, and his father is still mentioned for the power which he wielded and for his many acts of impulsive generosity. In 1948 the Headman was the biggest landowner in Rampura, and it was rumoured that he had lent more than Rs 150,000 to people in at least 30 neighbouring villages. But the lineage of which the Headman was the leader was smaller than the lineage of which Nadu Gowda enjoyed undisputed leadership in 1948. Nadu Gowda was also more accessible than the Headman, but he was much less wealthy; even Millayya, a member of the same lineage, was wealthier. The Headman and Nadu Gowda were good friends and there was a great deal of understanding between them. Their friendship was partly responsible for Rampura's stability, a fact which was recognized by villagers who prophesied anarchy in the village 'after the two heads fall'. Between 1900 and 1920 the village was sharply divided into a few factions, with the present Headman's father and Nadu Gowda's father leading the two most important factions. Friendship between the present Headman and Nadu Gowda was formed in the teeth of their fathers' opposition.

During my stay in Rampura. I did not witness any dispute among Brahmins and therefore cannot say to whom they would have gone for ettlement. However, I do know that during a crisis Brahmins went to ne or other of the Peasant patrons for advice and help. When a caste s decisively dominant, its dominance extends over all the castes ncluding castes ritually higher. The caste-headman of Peasants in he neighbouring village of Kere told me that he once disciplined an rrogant Brahmin priest by imposing boycott on him. The high ritual osition which the Brahmin occupies does not free him from the ecular control of the dominant caste. This is also true with regard to he Lingayats. During the partition of the property of one segment of he Lingayat lineage which provides priests to the Madeshwara emple, the Headman and his sons were consulted by the head of the oint family, and the partition finally took place before Peasant arbirators. In a dispute between the priests of the Madeshwara and Basava temples, the Headman's advice was sought. This case is nteresting in showing the kind of issues on which the intervention of he dominant caste is sought. Soon after the harvest in 1948, the Basava priest, a widower, had gone east and brought back two loose vomen with him. This was criticized by everyone in the village. During the summer, at the annual feast in honour of Basava (*Basavana ara*), it is customary for the women of the priestly Lingayat families o join together and cook for all the Lingayats. These women began ooking and were joined by the mistresses of the Basava priest. The Lingayat women became annoyed and asked these women to keep way, as they were loose women and no one knew their caste. One of he Madeshwara priests later went to the Headman and requested im to see that henceforth the two priestly lineages cooked separately n such occasions. The Headman agreed. In 1952, in a case in which a Lingayat youth had a liaison with a Peasant widow and had also nsulted a few Peasant youths, the council consisted of the Headman, few other Peasants and a Lingayat priest. A fine was imposed on oth the parties to the liaison and it was decreed that the widow hould soon marry. The girl went to her sister's affinal village and got narried a few months later.

Even a group like the Muslims, with customs and traditions which re quite different from those of the Hindus, take intimate disputes mong close kindred to the Headman for settlement. In 1948 I ritnessed three such disputes between kinsfolk being taken to the Ieadman. It is the boast of Peasants in Rampura that the Muslims re unable to settle a dispute among themselves and have to take it to the Peasants. The Kere documents reveal a similar situation in Kere *hobli*.

I was told more than once that an effort was usually made to settle a dispute within the caste and to take it to the Peasants only when internal efforts failed. It took me some time to realize that this rule was more honoured in the breach than in the observance. I witnessed several castes taking their disputes, even intimate domestic disputes, to the Peasant patrons for settlement.

The Untouchables were the only caste to make an effort to settle their disputes among themselves. They even succeeded in recovering from a caste elder a fee which he failed to pay to the caste at the wedding of a daughter. I am unable to say whether the Untouchables took care of their own disputes because of new-found self-awareness as a group, or because they thought the higher castes would not be interested in their affairs.

Disputes between Untouchables and high caste men are taken to Peasants for settlement. Thus, in 1952, a dispute between a Lingayat landowner and his Untouchable agricultural servant was taken before the Headman. Inter-caste disputes are usually taken to the Peasants for settlement.

There is a marked tendency for disputes to be settled within the village. The local elders know the disputants intimately and they are more likely to take a sympathetic view than outsiders. Justice within the village is also cheaper, swifter, and more effective. The local elders either have direct power over the disputants themselves or have influence with those who have such power. This is why disputes tend to be referred to local patrons even across the caste lines. The power wielded by the local patrons is considerable, and even outsiders seek their intervention. Thus a Shepherd from a different village requested Nadu Gowda to use his influence with his wife's father in Rampura, to see that she joined him.

A man who takes a dispute that does not refer to caste matter outside the village is guilty of slighting the local patrons. His action is in effect, a declaration of 'no confidence' in them, and he will soon be made to realize that he has incurred their wrath. Nemesis is swift in an Indian village where people are bound to each other by a multitude of ties. The outside elders, on their part, would not like to offend local elders. They know full well the power wielded by the local elders, who would be able to withhold true evidence and even produce false evidene, if annoyed. Their help would be necessary in arranging match, in securing a loan, and in a dozen other ways.

In some cases, however, the local elders may not be likely to intervene. They may think that the particular question has to be decided by the elders of the caste concerned and not by themselves. In the summer of 1946, a caste dispute among Washermen from several villages in Mandya and Mysore Districts was settled in Rampura. None of the important Peasant patrons attended the meeting. Sometimes the Peasant patrons may be indifferent because none of the disputants is a client or a kinsman. When a client is involved in a dispute, the patron steps in either because he must, or because the client urges him to do so. Thus Nadu Gowda was actively advising a Shepherd client who was making efforts to get his daughter's marriage dissolved. The husband was living in Sathnur and the case had gone up before the Shepherds in that village.

Where the local patrons have power, and factions are not deep, disputes go before them for settlement. If they are taken to a government court or to elders living outside the village, it means that the matter is beyond the local patrons. Thus in Rampura, though people submitted their disputes to one or other of the Peasant patrons, there were a few who were known for wanting to take every dispute to the official courts. These people were not respected in the village. The Rampura patrons were rich and influential and it was their boast that their disputes were always settled locally. In this respect Rampura was unlike some neighbouring villages.

While there is usually a tendency to settle a dispute within the village, there is 'leap frogging' when the caste which is dominant in the higher village is the same as the one which is dominant in the lower village. Thus, if Peasants are dominant in both the higher and lower villages, the Peasants living in the lower village show a tendency to take every dispute, including trivial ones, to Peasant patrons in the higher village. This occurs even though it is the policy of the council of the higher village to support the authority of caste and village councils within its jurisdiction.

IV

A brief description of the structure of village and caste councils is necessary here. This is complicated, as a part of the picture has to be reconstructed from the little that is now open to observation. Elderly villagers are frequently heard to say that things have changed a great deal and that many of the customs and conventions which were being observed even 20 or 30 years ago are being dropped nowadays. The

Kere documents help a little in reconstructing the structure as it was a few decades ago, but it is not correct to argue that what was true of Kere was true of all *hoblis*. Moreover, elderly informants are fond of making neat statements about the social organization of their caste and area, and it is difficult to fit them to the behaviour seen today. Thus, for instance, a Potter will mention that at a wedding ceremony in his caste, 70 sets of betel leaves and arecanuts are kept apart for distribution to representatives from the 70 villlages forming part of Potter's caste circle. Another Potter will say that 48 or 60 sets ought to be kept and not 70; neither is able to list the villages. If one goes to a Potter wedding, one does not see the specified number of sets of betel leaves and arecanuts kept apart. One is told that the custom has been discontinued only in the last four or five years. The data are sometimes more reliable, especially when they relate to disputes which actually occurred and were witnessed by informants. With these strictures, I will endeavour to reconstruct the organization of village and caste councils.

The village council is the lowest unit in the settlement of disputes. In this connection it is necessary to define what is meant by a village. Where villages are nucleated it is not difficult to identify and distinguish a village, but every nucleated settlement is not regarded as a village by the government. Thus Kere consists of three distinct nucleated settlements, one of which is Kere proper; the other two which have distinct names, are called *dākhale grāmagalu* or 'satellite villges'. For official purposes, the two satellite villages are one with Kere, and the hereditary Kere officials look after all the three settlements. But for social and religious purposes, Kere is three separate villages. A small village is occasionally tacked onto a nearby larger one for reasons of administrative economy. In the settlement of disputes, however, it is the social and religious unit which is important, not the administrative unit.

For purposes of administration again, villages are grouped into *hoblis, hoblis* into *taluks* or *sub-taluks* and *taluks* into district, and finally, districts into the State of Mysore. Rampura is not only socially but also administratively a village, and it lies in Hogur Hobli, Sangama Taluk, Mysore District. (Unit August, 1938, Mandya and Mysore formed a single District.)

The division into *hoblis* is an old division, and a *hobli* may contain from 20 to 50 villages. This division corresponded to some extent with the social organizatioin. Thus the council of the capital (*kasba*) of the *hobli* was regarded as a *kattemane* or 'house of law'. A *kattemane* is a place where disputes are settled. The headman of the *kattemane* is called Nadu Gowda, and in the Rampura area he is usually from the dominant caste of Peasants. The area over which a *kattemane* has jurisdiction (identical with a *hobli*) is called Mahanadu (big country), and in letters, the headman of the *kattemane* is addressed as Chief of the Mahanadu (Mahanadu Gowda). It may be mentioned here that a basic dichotomy existed between the agricultural castes which constituted Nadu, and the artisan and trading castes which constituted desha. (These castes are represented on the brass ladle carried by the *kulavadi*, the Untouchable servant of the Hindu castes.) The agricultural castes are entitled to honorific suffix of 'gowda' and the trading and artisan castes to Shetti. The Chief of the Shetti group is usually a leading trader in the hobli capital and is called Desha Shetti. The Chief of the Nadu group of castes and the Desha Shetti are both respected figures, and sets of betel leaves and arecanuts are set apart for them at any wedding.

Peasants are dominant in Kere, and more than 50 years ago, they shared this dominance with Brahmins. But the Brahmins do not seem to have played a very prominent part in settlement of disputes. The Kere documents show that the council of Peasants wielded effective power not only in Kere village but over the entire *hobli*. I am not able to say what the situation was like in Hogur, the *hobli* to which Rampura belongs, but Rampura itself had powerful Peasant leaders, as we have seen.

A variety of disputes were referred to the council of Peasant elders in Kere. One document referred to the punishment of a Fisherman who falsely alleged that his father's classificatory younger brother's wife, i.e. his classificatory mother, was his mistress. If the allegation was true, both were guilty of incest. The Peasant elders of Kere were angered by the false allegation and felt much sympathy for the wronged woman; they imposed two fines on the man, one of Rs 40 to be paid to the Peasant council, and another of Rs 15 to be paid to the Fisherman council. The Peasant caste council threatened the culprit with expulsion from his own Fisherman caste if he repeated the allegation. This incident shows the extraordinary power possessed by the council of the dominant caste, being able even to threaten a member of a different caste with outcasting.

In another document the Peasant elders defined the conditions under which a Muslim priest resident in Ganjam village was to serve the Muslims of Kere. He was asked to recognize three Muslim in Kere as leaders of the local Muslims, and to serve only those whom these three approved, but told that if he were dissatisfied with the Muslim leaders, he was to take his complaint to the Peasant leaders. In the same document, the Muslims in Kere and another village agreed to obey the three Muslim leaders, and added that, in case of dissatisfaction with the leaders' decisions, they would appeal to the Peasant elders in Kere.

A year after this document was signed, Kere Muslims informed the Peasant elders that they would bring them their disputes, as some had refused to obey the Muslim leaders. This highlights a feature of rural social organization in this area: the council of the dominant caste tries to create a structure of authority within each group it has to deal with, though its efforts frequently fail. In one case the Peasant headman of a village in Kere hobli complained to the Kere council that his villagers did not respect him at all, but took every trivial dispute to the elders in Kere. He requested the elders to support his authority in the village. When the caste which is dominant in a village is also dominant in the *hobli*-capital, the many ties between the two groups seem to militate against settling disputes locally.

The word *kattemane* evokes respect in the minds of Peasants in this area. All castes resident in a *hobli*-capital claim that they have *kattemane* status. Thus Oilmen, Fishermen, and Muslims resident in Hogur claim that their councils are higher than their respective caste councils in the other villages of the *hobli*. An Oilman from Hogur, at a wedding in his caste in any other village in his *hobli*, demands that he be given a set of betal leaves and arecanuts (*veelya*) before any other Oilman from that *hobli*, and creates a furore if he is not given such priority.

Ceremonial precedence is a different matter from referring village disputes to the council of particular caste in the *hobli*-capital. If the particular caste is numerically strong and wealthy in the *hobli*-capital, disputes may be taken to its council. If one or more elders are also well-known for their wisdom and skill in settling disputes, more people use their caste council. In the course of a generation or two, such a place is likely to acquire renown for settling disputes. I am inclined to think that the reputation of the Peasant councils in villages such as Keragodu and Nelamane was built up in this way. It is also probable that over a period of time the reputation of a council rises and falls though people seems to regard it as immutable.

The council of the *hobli*-capital is sensitive about the position it enjoys. At the annual festival of Madeshwara in Gudi in 1947, a fight occurred between Kere and Bihalli, and in the month of February 1948 the elders of Bihalli began working toward a settlement. They approached the Peasant elders of Hogur and Rampura and it was decided to call a meeting of the elders of Kere, Bihalli, Hogur, and Rampura on a particular day. The Kere elders were informed accordingly, but no one from Kere turned up. Lame excuses were given for their non-appearance, but everyone knew that the real reason was that Kere people felt they were being ordered about by Hogur. On the next date that was decided, the Hogur elders failed to turn up. They wanted to pay Kere leaders tit for tat. The pride of both villages having been satisfied, everyone attended on the third day and the dispute was settled.

In actual discussions with villagers, they frequently mentioned a particular village as constituting the highest council (*andalu gadi*) for a particular caste. Thus Peasants stated that the Peasant council in Nelamane was their highest court, while Shepherds mentioned Chennapatna, Basket-makers designated Malavalli, and Potters, Keragodu. Informants always mentioned the existence of documents and copper-plate deeds which defined the rights and privileges of their caste. Of course, no one had seen those documents, but all had heard about their existence.

It seems likely that a particular caste dominant in a village gradually acquired a reputation for settling disputes; it is also probable that the elders at one time sought and obtained the support of the local chieftain, or ruler or the head of a monastery for their decisions. When the village is a *hobli*-capital, it is easier for it to establish and enhance such authority.

Evidence that the reputation of a council was not constant over a period of time is found on occasions when there is an open challenge to a dominant position. Thus, several years ago, Potters from Mysore City claimed precedence over Potters from Keragodu. The former argued that they represented the place where H. H. the Maharajah of Mysore lived, the capital of the State, and as such they had to be given priority over Keragodu. The Keragodu Potters replied that priority to Seringapatam had not been conceded when Hyder Ali and Tippu Sultan were ruling there (in the eighteenth century), and there was no reason why they should give precedence to Mysore now. Thereupon Mysore Potters refused to attend weddings in Keragodu area. However, the Mysoreans' claim was conceded in Tagadur area in the east. Refusal to acknowledge changes in relative status is the source of much confusion in regard to the organization of councils. The great increase in communications between members of the same caste living in different areas makes for more debate regarding mutual position, and there is yet another complicating factor: some young men in every caste regard all this as antiquated.

V

As already mentioned, the patron-client tie is of crucial importance in the settlement of disputes. It is so powerful that disputes are always referred upwards from clients to patrons.[3] As the patrons and clients frequently belong to different castes, there is no strong sentiment that a dispute should be settled within the caste. Disputes are more easily settled locally if the patrons are powerful and come from a caste which is decisively dominant. Where a caste is dominant in a group of neighbouring villages, the influence of patron extends far byond his own village. Where a village is split into factions, each faction administers justice within it, and interfactional relations resemble international relations.

In Rampura the biggest and most important patrons are from the Peasant caste, and the patrons from the other high castes are aware of the power wielded by the Peasant patrons.

In 1948, Rampura was not factionalized, though a few isolated groups disliked the Headman and everyone associated with him. They were not numerous enough to challenge his authority and that of Nadu Gowda. Elderly villagers, frequently remarked that Rampura was more prosperous and unified than in the first two decades of this century, when even the large landowners were indebted to money-lenders in Mysore and Tadagavadi, and when, during the harvest, grain was measured out to the creditors as interest on these loans. That earlier period was characterized by deep factionalism; the several patrons were at loggerheads with each other. Elderly informants asserted that the villagers went more to the law courts then than in the later period. Particularly Narase Gowda, a faction leader, went to the law courts a great deal and acquired an intimate knowledge of law and legal procedure which he put to excellent use harassing his enemies, including the present Headman's father. The

---

[3] This tendency is probably a universal feature of rural India. See, for example, Gough (1955).

two often supported rival candidates in law suits. Elderly villagers say that the factions in the village began to disappear sometime after the end of World War I when one of the leading villagers was involved in a murder case and the village closed ranks and rallied behind the suspected man.

I have heard it said in Rampura that respectable people ought not to frequent the law courts. In 1948 the few who did so were unpopular and had a reputation for being very unscrupulous. This does not mean that a respectable man should never go to a law court, but that he should go only after he has exhausted all other remedies. It is generally felt that it is better to settle a dispute in the village than take it to a government court. When in a moralizing mood, villagers are able to reel off the names of those who liquidated substantial fortunes in taking disputes from one law court to another. It may be mentioned that a good deal of what goes on in a law court does not make sense to villagers; they know that a clever lawyer has to be hired, and that when a man loses in a lower court he can appeal to a higher. When a man loses in a government law court it does not mean that he has done wrong or that he loses face with his fellow villagers, but only that his lawyer is not clever enough or that he is not lucky. Villagers know that a man who has a right to a thing may lose it in a law court and the man who has no right may win it. This contrasts with the decisions of a village court, which have an ethical connotation. For instance, I found in the village panchayat book a note to the effect that a Smith had been fined one pie (1/192 of a rupee). I asked some people about the meaning of fining an absurd sum like a pie. They explained to me that it was levied only on a man who was found to be a persistent wrong-doer over trifles. The fine meant that an eye had always to be kept on the wrong-doer. The more serious punishments such as imposing a heavy fine, temporary boycott, and outcasting also have an ethical implication. I do not mean to imply here that the decisions of village councils are always right and that village arbitrators are incorruptible, but only that the decisions have a moral implication which the decisions of the government civil courts usually lack.

I have mentioned earlier that the patrons of the dominant caste tend to support, if not to create, local structures of authority. In consonance with this principle, they apply to the disputants the customs and rules which the latter recognize as binding, even when they are different from the customs and rules which are binding on the dominant caste. This respect for the moral code of every caste is one of the reasons why the decisions of the council of the dominant caste still continue to be respected. It is indeed a matter of surprise that village councils continue to function in spite of more than 100 years of British law administered through the powerful official law courts. The Kere documents included two cases which had been pending before the Government law courts but which had subsequently been withdrawn to be submitted to the council of the dominant caste in Kere.

Summary

A study of the locally dominant caste and the kind of dominance it enjoys, is essential to the understanding of rural society in India. Numerical strength, economic and political power, ritual status, and Western education and occupations, are the most important elements of dominance. Usually the different elements of dominance are distributed among different castes in a village. When a caste enjoys all or most of the elements of dominance, it may be said to have decisive dominance.

The Peasants in Rampura enjoy decisive dominance. They command respect not only from several lower castes, but also from the priestly castes, Brahmins and Liṅgayats, who have a ritual rank but who are not free from the secular control of the dominant caste.

The numerical strength of a caste influences its relations with the other castes. The capacity to muster a number of able-bodied men for a fight, and reputation for aggressiveness, are relevant factors. Considerations of power do prevail. The members of the non-dominant castes may be abused, beaten, grossly underpaid, or their women required to gratify the sexual desires of the powerful men in the dominant caste. The patrons from the dominant caste are 'vote banks' for the politicians.

The dominant caste plays a very important role in the settlement of disputes, which are settled by the traditional village and caste councils and not by the modern statutory panchayats. A caste council usually has jurisdiction over only the members of a single caste, but the dominant caste has jurisdiction over all the castes living in a village. The leaders of the dominant caste not only settle disputes between members of different castes but are also frequently approached by non-dominant castes for the settlement of their internal, even domestic, disputes. In the settlement of disputes, the patron–client tie is extremely important.

A feature of the administrative system of Mysore, handed down from pre-British days, is the grouping of villages into *hoblis*. The council of the dominant caste in the *hobli*-capital is called a *kattemane* (house of law), and it settles disputes not only occurring within the capital, but also entertains appeals from councils of every village in the *hobli*. It normally tries to uphold the authority of local elders. The working of caste and village councils and their relation to the council of the *hobli*-capital is extremely complicated and perhaps varies from *hobli* to *hobli*. The study of the working of these councils is essential to the understanding of the dominant caste.

REFERENCES

GOUGH, KATHLEEN, 1955, 'The Social Structure of a Tanjore Village', in McKim Marriott (ed.), *Village India* (University of Chicago Press, Chicago).

SRINIVAS, M. N., 1955, 'The Social System of a Mysore Village', in McKim Marriott (ed.), *Village India* (University of Chicago Press, Chicago).

———, 1957, 'Caste in Modern India', *The Journal of Asian Studies*, 4: 529–48.

The Study of Disputes in an Indian Village

Within a few days of my arriving in Rampura village, I heard vague reports of a case of arson in which a poor man's straw-rick had been burnt down by a man from a neighbouring village. The better-off villagers each gave a head-load of straw to the injured man, and the result was that he obtained more straw than he had lost. The villagers did not give me the details of what had happened, and such facts as I obtained came my way a few months later when there occurred another case of arson. When a dispute occurs, people's memories are stimulated and precedents are quoted. Something like case law exists, though it is not systematized.

At about the same time, a widow brought a complaint against another woman who had accused her of leading an immoral life. I managed to secure a brief account of the incident but not at all in sufficient detail. It was clear that the villagers did not like giving information about the 'seamy' side of village life to an outsider. I felt that this was a challenge to me as a fieldworker. Besides, I must confess that like the villagers I found a dispute broke the monotony of village life, and gave people something to talk about. The villagers were quick to see the humorous side of disputes.

Disputes also had a dramatic quality. Thus one afternoon a man walked into my veranda dragging a lamb's skin with him, hurled it before Nadu Gowda, a respected elder, saying, 'Mrs Siddamma's dog ate up my lamb. You must secure justice for me.' Or again, another afternoon, Mrs Khasu, a Muslim, was pouring forth a Niagara of words in Kannada as well as Urdu, while laying her case before the Headman. The assembled men were all enjoying her oratory—in fact, some of them had previously expressed a hope that I would get a chance to listen to her oratory before I left the village. (There was a 'master' of abuse in the village, a peasant woman, and a boy offered to steal her fowl so that I could record her abuse!)

A good many disputes have a public as well as a private side. The former would take place in the field or street or on a veranda, while the latter, inside the house. Only in a few 'partition' cases was I able to witness a private session. The fact of my being kept out of the private sessions spurred me to devise ways and means by which I could get to know what had happened in them.

Every society has its own preoccupations, and whatever the problem the fieldworker is pursuing, he cannot entirely ignore the former. It is only in a village or area which has been already studied sufficiently intensively that he can ignore the preoccupations of the people to concentrate on his own particular problem. I was insensibly led into paying some attention to disputes even though my main interest was the delineation of intercaste relations. I am afraid that the amount of time and energy I could spare for disputes was not at all enough. This was especially so when I had to keep track, as I had to occasionally, of two or three disputes each of which ran on for a few weeks.

Partition disputes generally tend to drag on. When the idea is first mooted, it is at the end of a series of quarrels for which the women, especially those who have come in by marriages, are usually blamed. The elders who are approached to effect a division of the property among the coparceners usually advise them to stay together and keep their women in control. After a while quarrels break out again, and finally, the elders concede that it is better to divide than to quarrel perpetually. Then a second set of quarrels occur—how should the property be divided and who should get what? There are some conventions regarding this, but they do not prevent quarrels. After the property has been divided, one member feels that he has fared badly and he demands a redistribution. In such a case, adjustments are made with some difficulty and the document registered to ensure that similar demands are not made again. Another set of quarrels arises during the paddy-transplantation season when the bunds separating the flats are trimmed, and brothers, who are usually neighbours, accuse each other of encroachment. Rights of way across a brother's field and right to irrigation water flowing through it, are other matters over which disputes occur. Such disputes go on for years. The partition of property among brothers does not promote amity and it is frequently found that adult brothers are not on speaking terms with each other. While the members of a lineage show solidarity among themselves in relation to other lineages, there are tensions. The narrower the lineage-span, the greater the tension. An exception to this rule is the elementary family when the siblings are still very young.

Besides the reluctance of the people to discuss the seamy side of their life before a respected outsider, there are other difficulties. Only some of the 'facts' of a dispute are accepted as such by all. And even the 'same' facts are fitted into different configurations by different people. The arbitrators as well as the neighbours and onlookers know the disputants intimately, and everyone has his own image of the character and personality of each disputant. This is a pre-existing image and the facts of the dispute are woven into it. But the image is not unchangeable.

Let me give an example: in a dispute between two Oilmen who were uterine brothers, the elder brother's wife, a strong personality and an attractive woman, was found walking in the direction of the river Kaveri at about 3 pm. A farmer saw her and asked her where she was going and she replied that she was going to the river. She was so fed up that she wanted to drown herself in the river. When this was mentioned during the dispute, a few men laughed and said, 'Is she the type that commits suicide?' One of those who laughed was an arbitrator. Here the 'objective' fact is, the woman walking to the river and expressing her intention to drown herself. This is interpreted differently by different people. The danger is that interpretation and fact are so closely woven that if the sociologist is not continually on the alert, he is in danger of accepting some interpretations as facts.

These interpretations are not haphazard but are related to other factors. Thus, a decision of the village or caste council is often explained by saying that the Headman or another powerful arbitrator wanted to favour his kinsman or casteman or friend or client. In 'The Case of the Potter and the Priest' the Headman was stated to have changed his decision overnight about the punishment to be meted out to two people accused of fornication because an agnatic kinsman of his, accused of attempted rape (see pp. 175–8), was suddenly brought before him. He could not pass a harsh sentence on one and a lenient one on the other. The interests of a powerful man like the Headman spread everywhere and he is likely to remember his interests while judging cases. An arbitrator also has his prejudices. Thus Nadu Gowda, normally a fair man, disliked one Untoucbable in particular, and this came out sharply whenever a matter concerning him came up for discussion. Friendships are common and occasionally cut across caste lines, and they influence the interpretation of events by witnesses as well as arbitrators. Finally, the solidarity of the dominant caste and the kind of local leadership which it has, are relevant facts in the dispensation of justice.

Where the defendant is a powerful leader of a large faction, the arbitrators tend to be lenient because the defendant is capable of defying them and thus endangering the entire fabric of village law and order. (I am assuming here that factions are not so deep that the village council no longer functions.) There are saws which elders quote: 'We floated the matter away' (winked at it), 'We let it slip through our fingers' (ignored inconvenient facts), and so on. One arbitrator mentioned how when he raised a point during the settlement of a dispute, the headman's son winked at him to make him keep quiet. The poorer villagers are heard complaining about the corruption of arbitrators.

I must hasten to add here that this does not mean that the arbitrators can do just what they like. The ideal of justice (*nyāya, dharma*) is there, supported by moral and religious sanctions. The arbitrators cannot entirely and consistently ignore public opinion. There are also unwritten rules of evidence. In 'A Caste Dispute among the Washermen of Mysore', the defendant trapped the plaintiff by making him eat food handled by her, and also took care to see that a witness was present on the occasion (see page 130). This was one of the crucial facts in the case. As I mentioned earlier, one of the tasks of village councils is to determine what are the facts of the case. Evidence is insisted upon, and a distinction is made between direct and hearsay evidence. The reputation of a witness is important in evaluating the truth or otherwise of his statements A person is sometimes made to swear to the truth of a statement in a temple. But this is an extreme measure.

The tutoring of witnesses seems to occur frequently and this makes the arbitrator's task all the more difficult. In some cases, tutoring is not necessary as the man has an interest in *suppresio veri* and *suggestio falsi*.

It is usual for a man to know only some of the events which have occurred, but he maintains that what he knows is not only true but is the whole truth. This was brought home to me when I was taking notes of the dispute between the Potter and the Priest. What I did then was to confront one informant with another's version. It is obvious that several versions are more likely to yield the truth than a single version.

Then there are men who have a vested interest in disputes. They try to further their interest which may be monetary gain, or a trip to town ostensibly to see a lawyer or off :ial, and so further their sense of self-importance. The existence of such men is not only recognized, but they are credited with having even more influence than they actually do. (They also provide convenient scapegoats.) The words 'chitāvani' (instigation) and 'kitāpathi' (love of creating quarrels) are frequently heard in the village. My invaluable friend and assistant Kulle Gowda was active whenever a dispute occurred. His capacity for making mischief was widely recognized.

Once the sociologist has obtained an idea of the prevalent pattern of antagonisms in the village, he can use this knowledge for obtaining better information. Thus the friends of a man will provide one version of events while his enemies provide another version. And there are a number of marginal people who may provide a third version.

For a period of two years after leaving Rampura I was unable to so much as glance at my field-notes. When I finally came round to writing up a few disputes for a fieldwork class I experienced a certain amount of difficulty in achieving a completely coherent account. This was specially so with the partition disputes which usually ran for a few weeks and involved much acrimonious discussion. Some of my entries were vague or mutually inconsistent, and in the process of producing a coherent account of the dispute, I had to omit, change and reinterpret some parts of my notes. I mentioned this fact in my first pubished account of a dispute (see pp. 146–7).

Social anthropologists have in recent years stressed the fact that their descriptive monographs are a contribution to history. They claim that these monographs provide better data for future historians of primitive and peasant life in different parts of the world, than are available for any country and for any period in the past. This is no doubt true but it is essential to state that a social anthropologist's note books occasionally contain entries that are wrong, vague or partial. This is specially true of the data collected in the first few months. When he is writing up, the social anthropologist discards the entries which he knows or suspects to be wrong. But he rarely mentions that his clear and coherent accounts of various aspects of the life of the people he has studied are occasionally produced from notes which are far from clear and consistent. These difficulties exist in all cases excepting where the fieldworker has periodically taken time off from the field to read and ponder over his entries and resolves his doubts and difficulties by discussing them with his informants. They are particularly prone to occur where the fieldworker is spending a year or less in the field and also when he is recording disputes which occur over several weeks. I am not concerned here with the other limitations of fieldnotes as historical documents, namely, the subjectivism imposed by the fieldworker's interests, his limited energy and the degree of his conscientiousness. It is obvious that where the social anthropologist uses an interpreter, as he frequently does, the notes do not have the same value as when he has enough mastery over the language of the people he is studying.

Recent research has shown that even the genealogies recorded by an anthropologist do not always provide an accurate record of descent. This is especially so in segmentary societies where the genealogies regularly adjust themselves to the dynamics of the lineage system (Evans-Pritchard 1940: 216). Even where there is a caste of genealogists whose business it is to record genealogies and bring them up to date periodically, they do not always provide an accurate record of descent at all levels (Shah and Shroff 1959: 40–70). Generally speaking, the remoter the past, the less reliable are the memories of informants. Even with regard to events which happened a year or two ago, informants' memories are not particularly reliable. But where a large number of people are involved, several can be questioned to obtain an account which is broadly true. And where documents exist, informants can be questioned on the basis of the documents. I used the first technique in gathering facts about a dispute which had occurred in October 1947 between Kere and Bihalli, and the second in my account of the Washerman dispute.

It was while collecting the facts of the dispute among Washermen that the idea occurred to me to look for documents referring to settlement of past disputes. I was told that caste and village headmen in the big villages had such documents. I had no luck, however, with the Peasant Headman of Hogur, the *hobli* to which Rampura belongs, but I fared better with the Peasant headman of Kere, at a distance of three miles from Rampura. In the summer of 1952, I made several trips to him and finally obtained loan of over seventy documents, some of which referred to settlement of disputes which had occurred in Kere Hobli during the years 1900–40. The documents referred to a wide variety of matters, and I am convinced that where such documents exist, they are invaluable for the study of rural social history. My own analysis of the concept of the dominant caste owed much to these documents. I do know that such documents also exist elsewhere. The people with whom these documents exist do not take enough care to preserve them, and white ants, cockroaches and the monsoon are steadily diminishing the quantity of documents available to the anthropologist. These remarks also apply to village records lying in the *taluk* offices everywhere. Documents [such as these] have, somehow, failed to attract the attention of historians in spite of their obvious importance.

The systematic study of disputes in rural areas and their settlement by non-official *panchayats* constitute an important field of research. It is completely neglected at the present moment by sociologists as well as lawyers. The latter confine themselves to laws passed by the State and Central legislatures. Customary law as observed in the villages is not regarded as law even though it governs the lives of millions. Convenient myths exist to the effect that the introduction of British law destroyed the law and customs followed by the village *panchayats*. Indian villagers are really 'bilegal' using both their traditional system as well as British-introduced law administered by the official courts situated in towns. I have been told of cases withdrawn from the latter to be settled before the unofficial *panchayats*. The study of the effects of introduction of British law on the indigenous system and on Indian society needs to be investigated by historians, anthropologists and lawyers.

The concentration on formal and written law has distorted the perspective of Indian lawyers and intellectuals. It has led to even pretending that the law enforced in the unofficial *panchayats* is not law.

I am convinced, however, that the study of the submerged legal system is extremely important and will be one of the things which will have to be undertaken if we plan to develop a much-neglected field of study, namely, the sociology of law and legal institutions. Such a study will also throw light on a historico-legal riddle, the relation between law as embodied in the sacred books of the Hindus and law as actually observed and obeyed by the bulk of the people living in villages. Finally, the study of this problem is not unrelated to the policy of devolution which finds much vocal support among modern India's leaders.

REFERENCES

EVANS-PRITCHARD, E. E., 1940, *The Nuer*. (Clarendon Press, Oxford).
SHAH, A. M. and R. G. SHROFF, 1959, 'A Caste of Genealogists and Mythographers: The Vahivancha Barots of Gujarat', in Milton Singer (ed.), *Traditional India: Structure and Change (American Folk Society*, Philadelphia, pp. 40–70).

A Caste Dispute among the
Washermen of Mysore

The Dispute

I

The dispute which I have described in the following pages, occurred among the members of the Washerman Caste (Madivala Shetti or Agasa)[1] in Mandya and Mysore Districts of Karnataka State during the year 1946. It began formally on 20 May 1946, in Bihalli, at a wedding, when Maya of Bella, the plaintiff, charged the defendant, Shiva of Magga, with having given his niece Javni in marriage to Kala, the son of Kempi of Kotti, the sister of Arasi who had been outcasted about twelve years previously for having had sexual relations with a Holeya (Untouchable caste of Karnataka). (See Document III and the genealogical chart).[2] Maya's charge needs to be explained! According to it, it was not Kempi, but her elder sister Arasi who was outcasted. The guilt extended to Kempi only by association or identification with Arasi. This is a crucial point and will be discussed later. The defendant Shiva was brought within the orbit of guilt because he gave his younger sister's daughter in marriage to Kempi's son, which again provides an instance of guilt by association. It is necessary to add here that normally it is the parents who choose a partner for their son or daughter, and as between the two parents the father's responsibility is much greater than the mother's. It is only when the father is dead or very ill that the father's or mother's relatives, especially the former, have the responsibility of finding a spouse. In the above case it has to be presumed that Shiva's sister was either a widow or a divorcée, and that her husband's brothers had no

---

[1] The common Kannada word for a Washerman is Agasa, which is a caste name, the more high-sounding Madivala Shetti, is only used in the rural areas, and that too not commonly. The term Shetti is a honorific title used at the end of either a caste or personal name by some artisan and trading castes such as Washerman, Potter or Trader. It is used on formal occasions, or in documents. In the present documents, the suffix, Shetti, after a personal name, is used somewhat erratically.

[2] Henceforward capital D will be used for Document. The English version of the documents is given at the end of the text.

A CASTE DISPUTE AMONG THE WASHERMEN
OF MYSORE

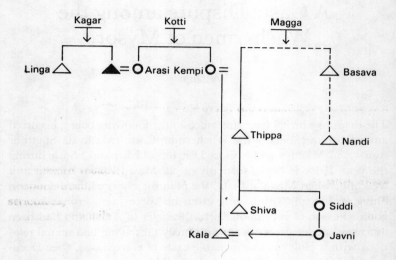

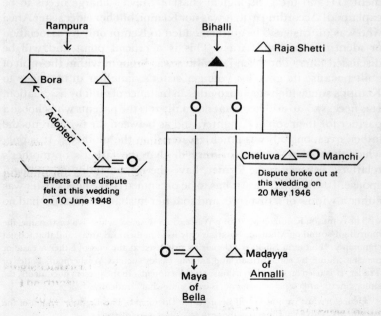

LEGEND: Underlined names refer to villages.

influence upon her. Perhaps she had gone to live with her brother after her husband's death, or after divorcing her husband. This is the only explanation for Shiva's having the power of disposal over his niece.

The timing of the charge was significant. The elders of the Washerman caste from several neighbouring villages had gathered together for the wedding at Bihalli, and Maya gave utterance to his charge when the guests were about to sit for dinner. Maya was asked whether he could prove, or provide evidence (*rujuvatu*) for his charge. He gave an undertaking in writing (*muchaḷikē*) that he would do so (D III). In D I the defendant Shiva agreed to subject himself to such punishment as the members of his caste (*kulastharu*) thought fit, if Maya succeeded in proving his charge (*phiryād*) before witnesses to that document. It was reported that Shiva was made to leave the dinner pending the inquiry. This was an insult and the Bihalli hosts were annoyed with Maya for causing this unpleasantness at an occasion like a wedding.

D II is a letter written by the plaintiff on 1 June 1946 to an elder in Rampura informing the latter that in accordance with the decision arrived at the wedding, a meeting of elders to adjudicate the dispute had been fixed for Tuesday, 4 June 1946, at the Mādeshwāraswāmi temple in Gudi, a village about a mile from Rampura. (The time of the meeting was mentioned as 10 a.m.) The plaintiff informed the Rampura elder that the responsibility for producing the defendant before the panchayat was his (the elder's). It is interesting to note that the dates mentioned are according to the Gregorian calendar, and not the Hindu calendar which is still used for all ritual occasions.

The inquiry into the dispute began on Tuesday, 4 June 1946 at the temple[3] mentioned above, before a large body of assembled elders of the Washerman caste. One or two informants estimated that several hundred Washermen had gathered on that day. Bora of Rampura, one of the hosts, had cooked 35 *seer*[4] of rice for his friends and supporters alone. Kempi, who was accused by Maya of having been outcasted, was made to serve food to these guests—this was a tactical move on the part of her friends, aimed at proving that everyone had social relations with Kempi and that she had not been thrown out of caste.

[3] It is probable that the meeting was actually held in the shade of the sacred peepul tree by the side of the temple.

[4] A *seer* measure is equal to about 2½ lbs.

In D III Maya mentioned both the immediate and ultimate origins of the dispute. The immediate origin was at the wedding in Bihalli, and the ultimate origin was the outcasting of Arasi, the elder sister of Kempi. Maya stated that ever since Arasi's outcasting, no member of the Washerman caste had any social relations with her or her sister. (Social relations include inter-dining, marriage, and other forms of interaction which normally prevail among the members of a caste). Maya challenged that if he was proved wrong he would pay whatever damages the elders thought fit. He actually used the word "damages"—it was spelt *dāmij* in Kannada.

D IV is a statement of the defendant Shiva. It is written in a somewhat formal manner clearly imitative of the records of proceedings in urban courts of law—some of the participants, including Kulle Gowda of Rampura, were familiar with lawyers and law courts in the towns.

Shiva answered Maya's charge by stating that for eleven years since the outcasting of Arasi he had no contact with either her or Kempi. But about a year and a half ago Maya (the plaintiff) had himself gone to Shiva and told him, 'You must see that your younger sister's daughter is given in marriage to Kempi's son.' Shiva replied, 'I agree, as you have done me the honour of coming and asking me.' Sometime later Shiva and Kempi went to Bella to obtain the advice of Maya regarding the arrangements for the wedding. Maya replied that he had no time that day and that he would come another day. He also suggested that the wedding should be held a month later. Shiva and Kempi agreed to Maya's suggestion, but when, some days later, they went to invite Maya to the wedding, he said, 'The marriage is between the bride's people and the groom's people. I have nothing to do with it. I will not come.' The marriage was celebrated without Maya. Shiva added that all his relatives continued to have social relations with him and Kempi.

In brief, the line of defence adopted by Shiva was to make Maya responsible for his (the former's) having social relations with Kempi. He also stated that neither he nor Kempi had been outcasted.

It was possible, however, for Shiva to adopt a different line of defence: he could have separated the charges against Arasi and Kempi, and argued that the outcasting of the former did not extend to the latter. Perhaps Shiva felt it was safer to steer clear of the career of a strong personality like Kempi. But, as will become clear later, the charge against Arasi did get separated from the charge against Kempi in spite of Maya's attempt, although not explicit, to mix up the two, and to condemn Kempi by implication.

D V is Kempi's statement of her defence. It is evident that this was made after the witnesses for Maya had given their evidence. Kempi roundly declared that everything that had been said on behalf of Maya was false. Another statement of hers cut the ground under Maya's feet by separating the charges against her and those against her elder sister. She stated that for a long time there had been no social relationship between her and Arasi. She added that Maya and some witnesses on his behalf, such as Elehalli Lingappa and Boodnoor Mallayya, had been having social relations with her all along. Finally, like Shiva, she made Maya responsible for the marriage between her son and Shiva's niece. Her version of the wedding negotiations was substantially the same as Shiva's. There were minor discrepancies no doubt, but these were not picked on by the opposition to puncture the evidence of either Shiva or Kempi. Perhaps the discrepancies were not there in the statements as made but only in the written record. But the documents have to be taken, in spite of their great brevity, and obvious deficiencies in draftsmanship, as giving a fairly accurate account of what happened. The great evidential value ascribed to documents in rural India ensures that what is reduced to writing in the presence of several witnesses representing different 'parties', is essentially true. The possibility of deliberate cheating by a clever and corrupt scribe should not, however, be ruled out.

D VI is a statement by Boodnoor Mallayya, described by Kempi as a 'witness for Maya', but at head of D VI it has been mentioned that he is a witness for Kempi! His evidence was quite crucial as is recognized in D XIV: it is mentioned there that it was on the strength of his evidence that the case against Shiva and Kempi fell through, and eventually, it led to a fine being levied on Maya. Mallayya's statement was regarded by the panchayat as providing a clear evidence that Maya had continued to have social relations with Kempi himself, which clearly disproved his allegations that she had been outcasted along with her sister, Arasi. Or, alternatively, if Kempi had really been outcasted, and Maya had eaten food touched by her, he too had been rendered impure.

Mallayya told the panchayat that some time previously he and Maya had occasion to return to Bella from Bihalli *via* Kotti, Kempi's village. Maya took Mallayya to Kempi's house—this was Mallayya's first visit to her house. After a little talk, Kempi told the guests that she had some steamed cakes bought for them from the shop, and that she would make some coffee. She requested the guests to wait till the coffee was made.

When Kempi went in to make coffee, Maya and Mallayya went towards the village pond to perform ablutions. I was, however, told by some informants that during the period of Kempi's absence in the kitchen, Maya came to learn—it is not known how—that Shiva and Kempi have become lovers. This made Maya very angry indeed and he left the house at once without partaking of Kempi's hospitality.

Kempi must have suspected that something had gone wrong when she found that her guests had suddenly disappeared without informing her. But, clever woman that she was, she carried the food on her head to the two men, taking care to be accompanied by a man of the Okkaliga (Peasant) caste, a different and higher one than the Washerman, to act, if and when necessary, as a witness. Kempi's actions point to premeditation on her part, and but for it she would not have been able to outmanoeuvre Maya.

To Kempi's and the Okkaliga's queries as to why they had left so suddenly and without informing anyone, Mallayya replied that they wanted to perform ablutions. After a while the two guests ate the food brought by Kempi in the presence of the Okkaliga. The entire incident suggests that Kempi was able to anticipate Maya's reaction when he learnt that Shiva had become her lover.

Mallayya was followed by Annalli Madayya (D VII) who began his evidence by declaring that he was Maya's (classificatory) brother. His evidence confirmed Mallayya's: he said that he had been having social relations with Kempi for the past five years and that he and Maya had occasionally dined together at Kempi's house. He also deposed that there was harmony among them all until the incident at the wedding. He had seen Washermen from Bihalli and other villages partaking of the dinner served at the wedding of Kempi's son and Shiva's niece.

A point which is worth commenting upon is that in D XIV Mallayya's evidence is referred to, but not Annalli Madayya's, in spite of the latter being the classificatory brother of Maya. Whether this is merely an omission, one of several, or whether Mallayya's evidence was given greater weight because of the presence of an impartial third party viz., the Okkaliga companion, it is not clear.

D VIII is a brief statement by one Rudra of Settalli saying that he had been having social relations with Kempi for the past six months and was unaware of anything else in the case. This statement, along with Ds VI and VII, aimed at proving that Kempi did have social relations with her castefolk. From which it followed that the ban against Arasi did not extend to Kempi.

D IX only confirms the evidence of the three earlier documents. In it Kempayya of Bella declared that while everything he knew about Arasi was direct and first-hand, what he knew about Kempi was only hearsay. He added that he was eighty years of age. I was told by informants that Kempayya's evidence carried considerable weight because of his advanced age and reputation for honesty. His clear separation of direct '*record*' from hearsay (*dubāri*) evidence was also appreciated. D XIV does not, however, contain any reference to Kempayya's evidence.

Ds X–XII are all statements by witnesses for the plaintiff, Maya, D X is a statement by Kyatayya, an elder of the Washerman caste in Kotti, Kempi's village. He alleged that sometime ago the Washerman elders of Kotti had received the information that Kempi had resumed social relations with her outcaste sister, Arasi, after having broken them when the latter was expelled from caste. The elders sent for Kempi who was, according to Kyatayya, in a defiant mood. She is alleged to have told the elders that she preferred her sister to her caste. A ban was then imposed on her. Since then no one, with the exception of Shiva, had any social relations with Kempi.

Ds XI and XII are both statements of Linga, the younger brother of Arasi's dead husband. In D XI he provided evidence in support of the allegation that Kempi did maintain social relations with her elder sister and in D XII he adduced evidence on the ultimate source of the present dispute viz., the outcasting of Arasi for having had sexual relations with an Untouchable.

Linga stated that he and Arasi occupied two partitioned halves of a single house—when a partition takes place all joint family property, immovable as well as movable, is divided, and quite frequently, a single house is divided by walls into two, three or more parts according to the number of sharers. In such a case each partitioned household gets to know a good deal of what goes on in the other households. It is necessary to make clear here that Linga began by stressing the advantages of his position for observing the conduct of his dead brother's widow. He stated that whenever Kempi visited Arasi, the two had dinner together. He reported the matter to the panchayat who then issued a rule (*vidhi*) to the effect that none may have social relations with Kempi. In D XII Linga deposed that about twelve years ago he reported to the elders of his caste the fact that his elder brother's widow had sexual relations with an Untouchable. He caught the pair redhanded. The elders took a statement from Arasi before 'releasing her from caste' i.e. outcasting her. Since then all social relations with her were suspended.

In D XIII a parallel cousin of Shiva gave evidence against the latter. According to him, Kempi had approached the elders of the Washerman caste in the previous year in Magga, with a request that they should help her to secure Annalli Madayya's daughter for her son. Shiva's father reported that Kempi maintained social relations with her outcaste sister, and she was made to leave the panchayat immediately. Shiva was asked whether he maintained social relations with Kempi to which he replied in the affirmative.

It may be pointed out that pollution is the means by which guilt spreads. Pollution may be described as social contagion. Whoever comes in contact with the guilty party also becomes guilty. Marriage, sex and dining constitute contact always, and in certain situations, contact may have a wider connotation.

D XIII has appended to it the signaturee of five Washerman elders in addition to that of Shiva. In D XII(a) they declare that they have seen certain things with their own eyes and reflected with their minds on what they have seen, and they have come to the conclusion that *both* Arasi and Kempi stood expelled from caste.

D XIV states the verdict of the assembled Washerman elders and others. Kulle Gowda of Rampura, who at that time held the purely titular and high-sounding office of Village Organizer, was present as the document in question eloquently testifies. The reference to the Nadu Gowdas who are officials of the traditional social structure in this part of Mysore State, and who come from all the high castes in the area, illustrates how the decision of a caste court is supported by officials representing the entire society. Caste cannot exist by itself— it must take note of the village and of the wider territorial units.

In D XIV Maya is described as admitting to eating cakes and drinking coffee given by Kempi. This was witnessed by Boodnoor Mallayya. Maya was fined the traditional sum of 12 *hanas* (Rs 2), and the money was paid into the temple of Madeshwara. It is explicitly stated that the verdict was read out to the assembled elders who gave their approval. It is not, however, stated in the document that Kagare Linga (Ds XI and XII) was fined for giving false evidence (*sullu sākshi*).

II

A few facts which I was able to collect about the dispute and some leading personalities in it may not be out of place here. It must be mentioned that these facts were collected in Rampura in 1948 and 1952, two and six years respectively, after the dispute had occurred.

Kempi had the reputation of being both a virago and a loose woman. Even while the dispute was in progress she threatened to beat Maya with sandals—a great insult indeed, involving both the aggressor and the victim in a temporary loss of caste—and she said that she was prepared to spend Rs 2,000 on the ceremony of purification and readmission to caste (*kulashuddi*). The ceremony involved, among other things, giving a dinner to the castefolk.

Kempi's father seems to have had no sons. After his older daughter, Arasi, had married into Kagare village, he had his second daughter, Kempi, married in the *manevālathana* way. In this form, the son-in-law leaves his natal kin-group to join his conjugal kin-group, and the children born of the marriage are regarded as the children of the wife's natal group. Kempi had a son by her husband, and sometime later she became the mistress of Hori, a Peasant of Kotti village. Kempi's husband heard of this liaison, and he beat her. She complained to her lover who promptly took her away, and gave her shelter in the house of an Oilman (*Gaṇiga*). Kempi left her husband and son to join her lover. Though the two lived together, Hori did not eat food cooked by his mistress because he belonged to a higher caste. He cooked his own food. Kempi had a daughter by her lover, and two or three years later, Hori died. Her husband had predeceased her lover. She went back to her son who was about sixteen years old at that time. Maya became her lover soon after and he took an interest in the marriage of her son. Maya, an important elder of the Washerman caste in Bella. went to Shiva, his wife's sisters' husband,[5] and asked Shiva to see that his younger sister's daughter Javni was given in marriage to Kempi's son Kala. A little later Maya, to his chagrin, discovered that Shiva had become Kempi's lover. His annoyance when he made the discovery was so great that he refused to attend the wedding which had been arranged at his initiative. He then tried to get Shiva thrown out of caste for having social relations with Kempi, an outcaste—the very framing of the charge shows what a clever litigant Maya was. But, unfortunately for him, Kempi was more than a match for him.

[5] This relationship is not shown in the chart. I regret that I was not able to obtain the complete genealogies of all the parties to the dispute.

She anticipated his moves and trapped him successfully. It is surprising that Maya fell so easily into the trap. Either he allowed his desire for vengence to get the better of his sense of evidence, or he was confident of his ability to produce witnesses who would prove his charge against Shiva and Kempi.

I was told that Maya had paid Rs 100 to the members of the Washerman's caste-panchayat at Hogur. I do not know whether this is different from the bribe that was alleged to have been given to the five members who appended their signatures to D XIII. If it is the same, it is difficult to understand some of the facts mentioned below, assuming of course that they are true.

I was told by Bora of Rampura that as a result of this dispute, Keshava, the leader of the Washerman caste-panchayat at Hogur, was removed from his office for having accepted a bribe. (There is a problem here: Keshava's name does not occur among the five who signed D XIII.)

Bora's reliability as an informant should not be exaggerated, but I do know that he is not a *persona grata* with the Washerman elders of Hogur. In 1948, two years after the dispute had ended. a few days before the wedding of Bora's son[6] to a girl from Kollegal, the Hogur elders wrote a letter to the girl's parents saying that Bora had lost caste as a result of having eaten food cooked by Kempi. They advised the girl's parents not to give their daughter to Bora's son. Bora was agitated by this, and he took the documents of the dispute to show to his would-be affines. The latter advised him that he should file a suit for defamation in a court of law against the elders of Hogur. Here was an attempt to use the legal system introduced by the British to strengthen caste mores.

III

The subject of the dispute relates, in the last analysis, to the operation of the moral sanctions of caste. Maya tried to have Shiva outcasted because the latter had social relations with Kempi, whose sister had been outcasted twelve years earlier for having had sexual relations with an Untouchable. The sanction of outcasting is an extremely powerful one, especially in its more severe form which does not envisage a return to caste under any circumstance. In the milder form the offender is permitted to return after the lapse of a certain period, after he has performed the required expiatory rites and given his caste members a dinner and paid a fine. For instance, the offence of which

[6] Bora's younger brother's son, whom he had adopted.

Arasi had been found guilty was one which did not normally permit of readmission. On the other hand, dining, not deliberate, of course, with an outcaste, would not lead to the diner's being outcasted for good (see D XIII).

Traditionally, caste was able to exert great pressure on a member: even one who had defied it for a decade or two would try to return to the fold when he wanted to get his son or daughter married. Kempi, after a life of defiance, approached the elders of her caste when she wanted to get her son married (D XIII). Without the support of one's caste it is not possible even today for a person to obtain a spouse: I am not referring here to the high caste, Westernized groups living in the cities but to the millions living in the villages.

While a woman of a high caste would be outcasted for having sex relations with an Untouchable, a high caste man would not normally be thrown out of caste for having a *liaison* with an Untouchable woman. In the former case the panchayat would say that a mud pot defiled by a dog's touch must be thrown away, whereas in the latter they would say that a brass pot touched by a dog should not be thrown away but purified. Women are expected to observe a much sterner code of conduct than men.

Relations between members of different castes, and frequently enough, between members of the same caste, are governed by the concept of pollution. When all the members of a sub-caste in an area have social relations with each other, it implies that there is no pollution. When a member has been punished for an offence, having any relations with him, even drinking water from a tumbler touched by him results in pollution. Strictly speaking, there is no need for a deliberate act of punishment by the caste-panchayat—both the wrong-doer and castefolk consider that indulgence in a prohibited relationship results automatically in impurity. Readmission to caste occurs only after the necessary expiatory rites are performed.

In this dispute the Washermen of several villages were involved. In D III Maya issued a challenge that he would produce evidence of Shiva's guilt before all the leaders of the various caste-panchayats (*eltā gadiya kulada yajamānaru*); and again in D IV it is said that the Washermen of Mandya and Mysore Districts were present to settle the dispute. Several hundred Washermen are reported to have gathered together at the Madeshwara temple on that memorable occasion referred to earlier. This should provide some idea of the spread of caste ties—that too for a numerically insignificant caste like the Washerman—and the strength and vigour of caste as an institution.

Representatives of other castes too were present: it is stated that Maya paid his fine before the assembled Nadu Gowdas. The elders of a village or region frequently belong to different castes and recognize that each sub-caste has a sphere of action in which they do not interfere unless asked to do so. On the other hand, caste and village normally support each other.

IV

The effective unit of the caste system is not an all-India category like the Brahmin or Kshatriya, but a small local group such as the Kannada-speaking Washermen living in a number of adjoining villages. In other words, it is a small, homogeneous and well-knit group spreading over the villages of a small area. Usually the members of a caste in an area are all related to each other either as cognates or affines. The widespread prevalance of cross-cousin marriage frequently duplicates the kind of kinship bond prevailing between persons. So when a dispute breaks out, it disturbs the harmony of social relations among a closely-knit body of people. Existing cleavages deepen, new ones occur, and alliances are formed between individuals and groups. In the dispute discussed above, Shiva and his father were on opposite sides. Maya's mother's sister's son, Annalli Madayya, gave evidence against Maya. Maya and Shiva were themselves related as *shadgas*, viz., husbands of sisters. In short, existing configurations of relationships break down while a new one takes time to form. In the meanwhile there is conflict and confusion.

It is seen that caste disputes pervade the sphere of kinship. Even intimate relationships such as those of father and son, brother and brother, and brother and sister, are breached. Kempi was asked to suspend her relations with her elder sister Arasi. In D X she is reported as having said that she wanted her elder sister, not her castefolk. Whether this was true or not, it gives an insight into the nature of the conflicts produced by a dispute in a small local group the members of which are related to each other by a variety of ties. In such a case one generally obeys his caste, perhaps after an unsuccessful effort at defiance. As long as marriage has to take place within one's caste, the caste elders hold the trumps in their hands. And the sanction of boycott in a small face-to-face community is a very powerful weapon.

As has been said earlier, the procedure adopted in caste courts or panchayats is markedly influenced by the procedure obtaining in urban courts of law. In every village there are a few individuals who have journeyed many times to the law courts either on their own or on behalf of their kinsfolk or friends. Sometimes they act as touts for urban lawyers. They have a vested interest in disputes, and do not miss a chance to enlarge any existing differences. Village courts imitate only to a very limited extent the procedures of government law courts. In fact, not infrequently, documents which are parts of rural disputes carry the signature of irrelevant or wrong persons. For instance, there was no reason why D I should have carried Shiva's signature, or D IV, Maya's. Besides, the draftsman usually contributes his own errors and oddities usually in an effort to impress other villagers with his familiarity with lawyers and law courts. He introduces bombastic, technical terms and phrases, frequently wrongly spelt.

Where the procedure, however, differs radically from that of the urban law courts is in the choice of place for conducting the proceedings. In ordinary disputes, any odd veranda (*jagali*) before a house will do. I have heard an elder, sitting in his veranda, sharply reminding a loose-tongued woman, party to a dispute, that she was in a nyāya-sthāna (place of justice, court). But when a serious dispute, involving important people and several villages, is to be settled, the meeting is usually held before a temple. Every person begins and frequently ends his statement, by a reference to his 'soul as witness' (*ātma sākshi*), of 'god as witnesses' (*Ishwara sākshi*). He says that he is making the statement on oath (*pramāṇa*). A false statement under these circumstances is supposed to bring on the head of the guilty person, or on a member of his family, some disaster which is the result of divine wrath. But this does not mean that witnesses are always truthful—in the dispute first discussed, Kagare Linga was fined 3 *haṇas* for giving false evidence. The tutoring of witnesses is quite common.

The assembly of caste-elders is referred to as *Kulaswami* or 'lord caste'. It is treated with respect bordering upon reverence. A person prostrates before the assembled elders before being readmitted to caste.

Occasions such as a wedding when a number of castefolk come together seem to be chosen for bringing a complaint formally to the attention of the caste-assembly. Maya's behaviour, described in D III, may be said to be typical. When the wedding guests were about to sit for dinner, he announced that Shiva, himself one of the guests, had no right to be in the assembly as he had social relations with the outcaste Kempi. And he refused to sit for dinner till Shiva was sent out of the hall. Some of the members present must have demanded evidence for what he was saying, and consequently he gave an undertaking in writing that he would prove his charge. The date and place of the meeting were agreed upon on that occasion alone.

. It is probable that the case was settled in one sitting. Maya stated his charge formally (D III) and Shiva replied to him (D IV). As the crucial issue in this case was whether Kempi had been outcasted along with her sister, she was asked to make statement (D V). It is very likely that witnesses for Maya were called before witnesses for Shiva, but the statements of the latter have been recorded before those of the former. Kagare Linga made two statements. (Ds X and XI), the first one referring to Kempi's having social relations with her outcaste sister, and the second, to the ultimate source of the dispute, the outcasting of Arasi. The slightest doubt regarding the latter—or rather the failure to establish it as an indubitable legal fact—would have completely destroyed the case against Shiva. A perusal of the documents shows that the elders knew what really were the crucial issues in the dispute, and they were able to proceed straight to them in spite of a vast mass of obtruding and irrelevant data. In the present dispute, for instance, the question is, 'Is Shiva guilty or not of having had social relations with Kempi who is alleged to have been outcasted along with her sister Arasi?' The correct answer of this question involves finding out first of all whether Arasi had been thrown out of caste, secondly, whether Kempi had been expelled for continuing to have social relations with her sister, and finally, if the answer to the second question is 'yes', the further question arises, 'Did Shiva have social relations with Kempi?' Arasi was no doubt outcasted, but there was no evidence to prove that her sister too had come under the ban. On the contrary, there was evidence to show that she had social relations with the plaintiff. Shiva did not, therefore, commit any offence when he gave his niece in marriage to Kempi's son.

'Did such-and-such an event actually occur?' is a question frequently asked by a panchayat. How to establish the truth of an allegation before the panchayat? *Sākshi* or witness, *rujuvātu* or evidence, and *muchhalike* or written statement, are terms one frequently hears in the rural parts of Mysore. A panchayat will not accept that a particular event occurred unless evidence is provided for its occur rence, and what is evidence? Evidence can be of several kinds, according to the degree of reliability. But all evidence has this in common— it needs a witness. The character of the witness is a highly important fact—if he is a man of straw his words will not have value. A man of substance, possessing land and money, and with a reputation for piety and integrity, will sway the judges considerably. If, in addition, he is an elder, then the opposing party will have a difficult time controverting his evidence. As mentioined earlier, my informants thought that the evidence of Kempayya, the old man from Bella, as extremely important. He was known to be a man of integrity and he was eighty years of age. He made only a very brief statement saying that he knew everything about Arasi whereas he had no direct evidence at all about Kempi. The panchayat distinguishes between direct and hearsay evidence. The latter does not have much value. This helps to explain why a shrewd woman like Kempi took a man of a different and superior (also dominant) caste with her to witness Maya's eating of food handled by her. The Peasant witness was not called, but Boodnoor Mallayya's statement (D VI) was not contested by Maya.

There is also a difference between written and oral evidence. The former has more value than the latter. It is universal for joint family property to be partitioned before a few elders and the terms to be reduced to writing, but registering the deed is not so common. That a registered deed has even greater value than a merely unregistered one is recognized by rural folk. The weight attached to docouments is enormous even when the draftsman's acquaintance with the language is only elementary.

Normally the weight of evidence tells. In spite of the fact that Maya was a powerful man, and in spite of the alleged bribing of the judges, the decision was given in favour of Shiva. Maya was fined, and so was Kagare Linga, the latter for false evidence. The amounts of the fines are small, but what is important in punishment is not the financial loss inflicted but the loss of face.

The Documents

A Brief Description of the Documents

DOCUMENT I: The defendant agrees to abide by the decision of the caste assembly in the matter of the dispute.

DOCUMENT II: A letter written by the plaintiff informing one of the parties of the date and place of the meetings.

DOCUMENT III: Plaintiff's statement.

DOCUMENT IV: Defendant's statement.

DOCUMENT V: Kempi's statement.

DOCUMENTS: VI, VII, VIII and IX are statements by witnesses for Kempi, and ultimately, for the defendant.

DOCUMENTS: X, XI, XII, XIII (a) are statements on behalf of the Maya, the plaintiff.

DOCUMENT XIV: The court's decision.

DOCUMENT I

Agreement entered into on 20-5-1946 at the wedding of Cheluva s/o Raja Shetti of Bihalli, Hogur Hobli, Sangama Taluk. In the presence of the witnesses who have gathered here, I, Shiva s/o Thippa of Magga, agree willingly to pay whatever fine or other punishment my castefolk (*kulastharu*) may decide on, should the charge (*phiryād*) levelled against me by Maya s/o Bogayya of Bella, be proved true before witnesses.

Signed by SHIVA

Witnesses

(1) Ramu	(7) Harigolu Kempa
(2) Hogur Keshava	(8) Boodnoor Mallayya
(3) Kapi Nanja	(9) Rampura Bora
(4) Annalli Madayya	(10) Bettalli Veera
(5) Maghu Kempa	(11) Bihalli Kempa
(6) Settalli Rudra	(12) Kagare Linga

Document drafted by Kuri Malla.

[Washermen from fifteen villages are mentioned in the above document].

DOCUMENT II

On 1st June 1946, Maya wrote the following letter to Puttayya of Rampura: It was decided at Bihalli to arrange a meeting of castefolk a fortnight from 20th May. Accordingly, a meeting will be held on Tuesday, 4th June, in Gudi near the temple of Sri. Madeshwara. Please come to the meeting at 10 a.m.

Signed by MAYA

N.B. It is your responsibility to bring the defendant (*aparādhi*) Shiva s/o Thippa. Please bring him to the meeting at the time mentioned above.

[Temples are favourite places for the settlement of disputes. It is believed that people are less prone to perjury before a temple than elsewhere. Occasionally one of the parties may be asked to swear to the truth of a statement, and the settlement of a dispute is marked by *puja* being performed to the chief diety.

The Madeshwara temple at Gudi is a favourite place for settling disputes. In this case it was also the most convenient place, situated as it is in the centre of the villages involved in the dispute.]

DOCUMENT III

[A meeting was held on Tuesday, 4th June 1946][7]

Maya's statement before the meeting: I went to my brother-in-law's [*Cheluva's*] wedding at Bihalli on 20th May 1946. As soon as the *dhāré* was over, I mentioned [to the guests] that Shiva s/o Thippa of Magga has begun a relationship (*nentasthana*) with the house of Kempi of Kotti, sister of Arasi, who was outcasted for having sexual relations with an Untouchable (*holabālike*)'. I was asked to prove (*rujuvātu*) my allegation, and I agreed in writing (*muchhaliké*) to do so. I undertook to provide the evidence (*kaiphayat*) of relevant witnesses at an assembly of all the headmen of the caste-courts (*gaḍi*) of Madivalas to meet at the Madeshwara temple, 4th June 1946. About twelve years ago, after the decision against Arasi had been given, no one was normally having any relation (*bāḷiké*) with either Arasi or her younger sister Kempi. Food cooked by them was not eaten, nor were they allowed to dine with castefolk. If my statements are proved untrue, I am willing to pay such damages (*dāmij*) as you think fit. I have narrated the facts as they are before castefolk.

Signed by MAYA

[7] All statements in square brackets are my own. They do not occur in the documents.

Witness: Ramu

[1. *Dhāre'*: Sanskritic ritual at which the bride is given as a gift to the bridegroom. Also referred to as *kanyādana*.

2. Note the use of *dāmij* in the document.]

DOCUMENT IV

There was a meeting of the Madivala Shettis of Mandya and Mysore District at the temple of Madeswara in Gudi in Hogur Hobli, to inquire into the following dispute:

Plaintiff (*phiryādi*), Maya—Defendant (*aparādhi*), Shiva.

Defendant's statement: With my soul as my witness (*ātma sākshi*) I state firmly before the deity Parameshwara, and before the assembled caste-elders (*kulaswāmi*, literally, 'lord caste') that for eleven years (after Arasi's outcasting in 1934) I did not eat food cooked by Kempi, or have any other relation with her. About one-and-a-half years ago the plaintiff came to me and said, 'you must persuade your younger sister to give her daughter to Kempi's son'. I replied, 'I agree as you have done me the honour of coming and asking me'. Sometime later Kempi and I went to Bella in order to obtain the defendant's advice about arrangements for the wedding. The plaintiff told us, 'I have no time today. I will come another day. Have the *lagna* (wedding) fixed for next month, not for this month'. We fixed up the date of the marriage according to the plaintiff's instructions. Some days later Kempi and I went to the plaintiff's house to invite him to the wedding. He answered, 'The marriage is between the bride's people and the groom's people. I have no connection with it. I will not come to the wedding'. We celebrated the marriage. We continue to have all types of social relationship with our relatives. I have stated the facts as they are before the caste-elders.

Signed by MAYA and SHIVA

[1. It is likely that all statements by witnesses were made and recorded on 4 June 1946.

2. I learnt only in 1952 that Maya and Shiva had married sisters. I am not sure whether they were full or classificatory sisters.

3. Maya is addressed as *yajamān* (master or leader). I learnt in 1952 that before this dispute took place Maya was the leader of the Washermen in Bella and a few surrounding villages. The respectful way in which he is addressed by Shiva, and the readiness with which the latter agreed to give his sister's daughter to Kempi's son are indicative of Maya's importance. The attempts by Kempi and Shiva to ensure the presence of Maya at the wedding further strengthened this view.

4. Strictly speaking *lagna* means the time when the bride is ritually handed over to the groom. It is determined by the astrologer. In a loose sense it also means wedding.

5. It is not easy to understand why Maya signed this document; nor Shiva's signing Document XIII.]

DOCUMENT V

[The following statement by **Kempi** seems to have been actually made after the statements of witnesses of Maya].

I, Kempi, make the following statement before castefolk, with Iswara as witness: ever since the outcasting of my elder sister, there has been no social intercourse between her and me. All that has been said on behalf of the plaintiff is untrue. Maya, and witnesses on his behalf, Elehalli Lingappa and Boodnoor Mallayya, have beeen having social intercourse with me all along. Over a year ago I asked Maya to secure a girl for my son. He sent for Shiva of Magga, and told the latter, 'Arrange for your younger sister Siddi's daughter to be given (to Kempi's son)'. Shiva agreed. Subsequently he and I discussed the arrangements for the wedding and then went to Bella to consult Yajamān Maya. He said, 'I am not free today. Fix up the marriage in the month of Gouri (i.e. festival of Gouri, wife of God Shiva)'. When we went to him on the wedding day to invite him he said, 'I will come for the *dhāre* ritual'. But he did not come. I have stated the true facts before castefolk

Signed by KEMPI

[1. There is a little difference between Shiva's and Kempi's versions of what Maya said to them when they went to invite him for the wedding. See D IV.

2. Maya's position was high enough to send for Shiva and ask him to arrange for his niece to be given in marriage to Kempi's son.]

DOCUMENT VI

Statement of Malla Shetti s/o Boodnoor Doddayya on behalf of Kempi.

The plaintiff, headman, Maya, and I were returning from Bihalli to Bella *via* Kotti village, and Maya took me to Kempi's house in Kotti. Until then I had not gone to Kempi's house, nor had I dined there. After we had talked for a while Kempi said 'I have had some steamed cakes (*kadubu*) bought for you. I will make some coffee. Please stay here till I get the coffee ready.' Then we went towards the pond to answer calls of nature. In the meanwhile Kempi came to us, accompanied by a man of the Okkaliga caste, bringing with her steamed cakes, butter, jaggery and coffee. She sat her companion under a banyan tree, and asked us, 'Why did you leave? You ought to have eaten at my place before leaving'. Kempi's companion asked, 'Why were you trying to go away without eating the food offered to you?' I replied, 'We would not have gone away without eating. We came here to answer calls of nature'. Then I said (to Maya), 'They have brought food with them'. We ate the food sitting on the edge of the pond. Maya did not tell me anything about Kempi. I have stated the facts as they are before the caste-elders. I swear (I am telling the truth).

Signed by MALLA SHETTI

DOCUMENT VII

Statement of Annalli Madayya s/o Linga Shetti.

I am the plaintiff's mother's younger sister's son, and I have been having social relations with Kempi for the last five years. On her side she has been visiting our house occasionally. My elder brother (classificatory) Maya and I have both sat together (*saha pankti*, literary, same line) at Kempi's house, and eaten meals cooked by her. We have all been living in harmony (*eki bhāva*, literary, feeling of oneness). I do not know anything else. I have also seen Bihalli people and others eating in the wedding house.

Signed by MADAYYA

DOCUMENT VIII

Statement of Rudra s/o Rudra (Snr.)

For the last six months I have been having social relations with Kempi. I do not know anything else.

Signed by RUDRA

DOCUMENT IX

Statement of Kempayya s/o Kalayya of Bella:

I know everything (directly) about Arasi but about Kempi all that I know is hearsay (*dubāri*) and not direct (the actual expression used is 'not based on record'). I am about eighty years of age. I am telling the truth before castefolk.

Signed by KEMPAYYA

DOCUMENT X

[Kyatayya, whose evidence is given below, appears to be an elder of the Washerman caste in Kotti, Kempi's village.]

We learnt that Kempi had resumed relations with her elder sister after having severed them for some time. We sent for her and questioned her. She replied, 'I want my elder sister, I do not want you'. We then made a rule [*kattu*] that our caste people should not dine at Kempi's house. About a year ago, Shiva s/o Thippa of Magga, gave his younger sister's daughter in marriage to Kempi's son. Neither we, nor people related to us, have had social relations with Kempi. I make this statement with my soul as my witness.

Signed by KYATAYYA

DOCUMENT XI

[Documents XI and XII are both statements by Kagare Lingayya.]

I swear by God to tell the truth before castefolk. My house and Arasi's are but two halves of a single house. I used to observe that whenever Kempi visited her elder sister, she ate food with Arasi. But none of us went to Arasi's house. When the elders of the Washerman caste in Kotti told us, 'As Kempi is eating food cooked by Arasi, none of you may have social relations with Kempi', we decided to obey them. I am telling the truth before my caste-elders with my soul as my witness.

Signed by LINGAYYA

DOCUMENT XII

Arasi and I belong to the same group. That is, she is my *attigé* (elder brother's wife). About twelve years ago I saw with my own eyes my sister-in-law sleeping with an Untouchable (Holeya), and I reported the matter to the caste-elders. The elders took a statement from her and released her from caste. Since then we have not had social relations with Arasi. Even now neither we, nor our relatives have any social relations with her.

DOCUMENT XIII

Statement of Nandi s/o Basava of Magga.

I state on oath what I know before castefolk. About a year ago Kempi came to us to request us, in the presence of castefolk, to obtain the daughter of Annalli Madayya in marriage to her son. Thippa, my uncle (*Doddappa*, father's elder brother or mother's elder sister's husband), said, 'Kempi is having social intercourse with her elder sister who was outcasted for having slept with an Untouchable'. When   heard this, we made Kempi leave the caste-assembly. Then we learnt that our uncle's son Shiva was having social relations with Kempi. We questioned Shiva and he told us, 'I dine at Kempi's house'. Then we laid down a ban that Shiva may not have any social relations with us till he had performed expiatory rites.[8] I have narrated the truth before castefolk.

Signed by SIDDA SHETTI

(The signatures of the following five men appear below Shiva's.)

1. Borayya s/o Borayya (Snr.) of Doddakere village in Maddur Taluk.
2. Adayya s/o Adayya (Snr.) of Elehalli.
3. Chennayya s/o Chennayya (Snr.) of Saralu villave in Mandya Taluk.
4. Kariya s/o Kariya (Snr.) of Kere.
5. Siddappa s/o Siddappa (Snr.) of Oddur.

DOCUMENT XIII(a)

We the (abovementioned) five men have seen certain things with our own eyes and reflected with our minds on what we have seen, and we have concluded, like some others who have given evidence before, that both Arasi and Kempi have been expelled from caste. This we state on oath.

DOCUMENT XIV

This dispute (*nyāya*) was decided at Sri Madeshwaraswami temple in the presence of Rampura Village Organiser, Sri Kulle Gowdaru. Maya admitted to having eaten the steamed cakes and coffee given by Kempi. Boodnoor Mallayya witnessed the above act. As to Shiva's guilt, no evidence was forthcoming. Maya paid a (traditional) fine of twelve *hanas* into the temple before the elders of the *nān* (*nādu gowdaru*).

The dispute was settled on 4 June 1946 in the presence of Sri Kulle Gowdaru who read out the statement of Maya to the assembled elders of the Madivala Shetti caste. The elders approved of the verdict.

[1. Kulle Gowda was one of the arbitrators in Rampura. See

---

[8] The actual term used is *kulasiddi* which is a corruption of *kulashuddi*, i.e., caste purification. The member outcasted regains his caste, and the ritual is performed by a priest before the assembled castefolk.

my 'A Joint Family Dispute in a Mysore Village', for the part played by him in that dispute.

2. The sum paid by Maya is a traditional sum, and it is customary to pay the fine into a temple.

Kagare Lingayya (see Documents XI and XII) was fined three *hanas* (a hana=2 paise) for giving false evidence. His evidence may be the source of another dispute.

3. The term *nadu gowda* in the singular refers to a hereditary office held by a peasant elder, which enables him to settle caste disputes among members of his caste. In the plural, and in a looser sense, it refers to the important elders of the area.]

A Joint Family Dispute in a Mysore Village

I

I shall in this essay try and give an account of a partition dispute which took place in Rampura village during my stay there in the year 1948. I am treating the dispute primarily as a 'field incident'. I want to show how it gradually became less mystifying to me though even towards the end I felt that I had knowledge of only a segment of the total social reality which was the dispute; I want to show how, in the course of its development, the dispute gave birth to a number of minor disputes; and finally, I want to show the kind of evidence that is available to the fieldworker—a considerable part of it is hearsay, and it frequently consists of interpretations and evaluations of one person's words, actions, motives and personality, by another. The sociologist has in his turn to evaluate this kind of evidence and draw his conclusions from it.

I have presented here most of the information collected by me about the dispute. I have done it for several reasons, one of them being my desire to show the frequent change of point of view of the parties to the dispute and their protagonists. It is necessary to stress that there are always certain clear-sighted individuals in the village who are able to keep their interests in view all the time and they try to use every incident to further such interests. For instance, Kulle Gowda, who was one of the arbitrators, tried to use the dispute to work off a grudge he had against the village doctor. I have seen Kulle Gowda use the most unlikely situations to further his interests. While Kulle Gowda was an unusual kind of man, the use of situations to further particular interests of individuals went on all the time.

The total absence of privacy, and the fact that I was only spending about ten months in the village prevented me from attempting to write up systematically the notes I had jotted down everyday; and after leaving the village, preoccupation with other material prevented me from paying attention to the Rampura material till September 1950. When I went through my notes of this dispute with a view to writing it up, I discovered that several of the entries were only partially intelligible to me. This is not an unusual experience for a fieldworker, and I mention it only to make clear that I have had to reinterpret some of my own entries. This will give readers some idea of the distance between them and the actual events, many of which were first of all interpreted by the parties to the dispute and their protagonists. These were then recorded by me, always in haste, and not always totally accurately. This was the kind of raw data that was at my disposal for presenting an account of this dispute.

Studying Rampura is in some respects different from studying a primitive society without written history and writing—the peasants resort to a scribe whenever they want something to be recorded. Besides, the sociologist who studies an Indian village is not able to publish all the material he has collected as some of it might be plainly defamatory, while some other material, while not defamatory, might embarrass or give pain to many people. Even the device of altering the name of the village and of the chief participants, which has been adopted here, does not release all the material for publication.

II

The basic kin-group in Rampura was the joint family, though the elementary family was a close rival to it especially among the poorer people. But many of those elementary families had been formed a few years previously by the splitting of their parent joint families. This is only my impression, however. Where the elementary family is the basic kin-group, it frequently includes beside a man, his wife and children, an elderly parent or widowed sister of the man, or a relative of the wife.

The joint as well as elementary family may be looked upon as a traditional form. A joint family normally splits into a few elementary families in course of time, and an elementary family develops into a joint family when the sons marry and bring their wives home. But in some cases a family stays joint for a few generations because of the amount of land it owns, or its business interests, or its strong sense of family tradition and loyalty, or all these combined.

The joint family in Rampura is patrilineal and patrilocal. It consists of the descendants, in the male line, of a common ancestor, and their wives, sons, married as well as unmarried daughters. Sometimes after her marriage a girl leaves her natal home and becomes a member of her husband's joint family. The widow of a deceased member usually stays with her conjugal joint family except in the event of her remarriage.

The joint family is a corporate, property-owning, co-residential and commensal group. Every male member has a share in its property by virtue of birth in it, but he becomes entitled to sue for division only after he becomes a major. Such a member is called coparcener, and a joint family is a coparcenary.

The feeding and clothing of members, as well as their marriage and funeral expenses are met out of the joint family's income. All the male members work under the head of the joint family, who may be father or paternal uncle or eldest brother of the other members. The women work under the wife of the male head of the household though when the latter's mother is alive and active, she is the leader.

The partition dispute described here occurred in the joint family of the late Sadhu I. The male head of the joint family was Sadhu II, son of Sadhu I, and the female head, Kempamma, mother of Sadhu II and widow of Sadhu I. The joint family was made up of Kempamma, Sadhu II and his wife and children, Kempayya and his wife and daughter, Honnayya II, the minor son of the late Honnayya I, Kenchayya and his wife, and finally, Kiri, the youngest and unmarried borther of Sadhu II. The widow of Honnayya I had remarried and was living with her husband in Hotte, a village a few miles away from Rampura. Before the partition took place all the members lived in one house on the income from the ancestral estate.

A JOINT FAMILY DISPUTE IN A MYSORE VILLAGE

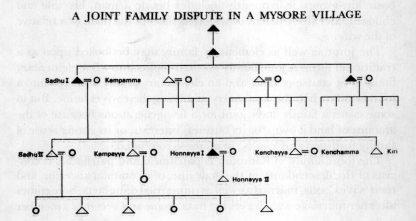

As the genealogical tree shows, the brothers of Sadhu I were all alive when the dispute occurred, and each of them was the head of a separate joint family. They all lived on the same street, and had to co-operate with each other on certain occasions like marriage and death though in the years preceding the dispute, differences had developed among the members of the lineage. I witnessed two disputes between these joint families during my stay in Rampura in 1948.

The joint families of the late Sadhu I and his brothers were all part of a still wider agnatic kin-group which was referred to as '*annatammike*', which may be transliterated as 'elder-brother—younger-brotherhood', and translated as 'brotherhood'. This brotherhood comprised thirteen joint families and included the wealthy and powerful joint family of the Patel or Headman of the village. The Headman was the most important man in Rampura, and by virtue of his office, wealth, the reputation of his lineage, and his personality, was the leader of the brotherhood. This brotherhood was not, however, the biggest brotherhood among the Peasants (Okkaligas) of Rampura. There was another brotherhood consisting of over thirty households, and considerable rivalry characterized the relationship between the two brotherhoods. Open conflict between them was, however, prevented by the great friendship which prevailed between the Headman and the chief of the other group, Nadu Gowda.

The brotherhood was an exogamous group. Whenever there was a birth or death in one of the houses of the brotherhood, all the members of it were affected by pollution for a certain number of days. They were also bound by devotion to a common *mane devaru* or 'house-deity' whom everyone within the brotherhood jointly propitiated periodically. The possession of a common house-deity was an index of agnatic relationship, and difference as to house-deity usually indicated an absence of such a relationship. Incidentally, the cult of the house-deity was found in southern Karnataka (formerly Mysore) State among Hindu castes.

III

The dispute in Sadhu's joint family was the first big dispute I witnessed in the village, and I came across it fortuitously on the night of Thursday, 10 March 1948. After dinner, I walked across to the Headman's veranda for a chat—every house in Rampura had one or two open verandas in the front, where strangers were received and friends sat down for gossip, and on which people slept at night, especially during the hot season. The verandas were well-known centres for gossip.

The Headman, his younger son Lakshamana, and Uddaiah, the eldest son of a former servant (and an agnatic kinsman), were sitting on the veranda while the disputants, five men and an old woman, were sitting in the middle of the street. (Those having a decidedly lower status than the Headman never sat near him, or on the same veranda). The faces of the disputants were not clearly distinguishable in the dark—the only light there was, came from the flickering flame of an earthen lamp which was in a niche high up in the front wall of the house. The disputants were arguing loudly and vigorously, constantly interrupting each other, and occasionally, interrupting Uddaiah, Lakshmana, and even the Headman himself.

I could follow the disputants only in bits and pieces. This was not because of any linguistic difficulty, but because they were speaking very fast and their talk was full of references to persons and incidents of which I knew nothing. The only thing I grasped clearly was that it was a joint family quarrel. I sat till 10 p.m. on the veranda and then returned home making a mental note to question Lakshmana or Uddaiah about the dispute the following morning. I learnt later that the disputants had argued before the Headman till midnight, after which a few of them had moved on to the immense log in the street in front of my house and continued arguing till well past 1 a.m.

In my first contact with the first big dispute I had witnessed, I had only heard a great deal of talk, and identified one of the parties (Kempayya, younger brother of Sadhu II). Everything else was a mystery.

IV

Friday, 11 March 1948. Three of the main characters in the previous evening's dispute came to my veranda in the morning. Lakshmana and Uddaiah, two of the three arbitrators. came about the same time. and Kempayya joined us later. I was sufficiently friendly with Lakshmana to ask him directly for an account of the previous evening's dispute. I also knew him sufficiently well to realize that his account would be far from 'objective'. Quite apart from his personal likes and dislikes, the fact of his membership of the Headman's joint family was bound to colour his approach to every village event. As already stated, the Headman's joint family was large and powerful, and its interests were all-pervasive. All the members of the Headman's joint family were remarkably loyal to it, and tried to further its interests in every context. *This fact should be remembered throughout.* Having uttered this necessary caution, I shall give Lakshmana's version

nan removes coconut beetles from trees. (Rampura 1948)

Castrating a bull. The man on the right is the local expert.
(Rampura 1948)

Village scene. The bridge across the canal provides a nice
spot to while away the time. (Rampura 1948)

ffic on the main Mysore-Bannur Road. The village tank is
he right. (Rampura 1948)

village tank shrinks into a pond. (Rampura, summer 1948)

of the events, and when I have any comments of my own, I will make them in square brackets.

Quarrels had occurred before in Sadhu's household and they had gone to arbitrators asking for advice. But of late, the quarrels were becoming very frequent.

The arbitrators normally counsel patience to those wanting partition. Village elders know the advantages of staying joint and the disadvantages of partition. For instance, when a family stays joint one earthen lamp suffices for the entire house. But when partition takes place, each commensal unit needs a lamp to itself. A man who is ill can stay in bed even during the transplantation season if he is member of a joint family, whereas he cannot do this if he has separated from his brothers, unless he has other relatives or servants to attend to his fields. Finally, partition frees the younger members from the healthy restraint of the elders. [Under the joint family system it is the head of the house who takes all the responsibility while the junior members do nothing but obey him in every matter. Not infrequently one comes across men over forty, occasionally grand-fathers, who look to the head of the joint family for every important decision.]

The girls who come into a joint family by marriage are the ones who are, in the last analysis, held responsible for the partition.[1] They do not like to be subordinate to their mother-in-law and to the wives of the husband's elder brothers. The brothers are after all sons of the same mother and even if they quarrel, they come together again. But their wives drive a wedge between them.

A man comes home tired and hungry from the fields in the evening. He wants a meal and then he wants to go to bed. But the women in the house will have quarrelled during the day and will be ready with their complaints when the men return. The house then has to become a court at sundown.

V

Lakshmana returned to the dispute after making the above general observations on the joint family. He said that Kenchayya was very fond of his young and pretty wife, Kenchamma. He also 'gave his ear' [listened] to her. [Both these are considered as not very desirable qualities in a young man. A man ought not listen to his wife, but to his parents and elder brothers. It is unseemly to exhibit one's affection for one's wife, while it is proper, even laudable, to profess a great loyalty to one's parents and brothers.] Kenchamma did not do her share of domestic duties properly—for instance, she did not thoroughly clean the house-front every morning, and she was able to flout the authority of her mother-in-law, because of her husband's support. Kempamma, the mother-in-law, was annoyed with Kenchamma's defiance of her orders, and so, one day, she beat her daughter-in-law with a broom. [Beating with a broom or sandal, or spitting on a person, render the latter impure. It is also considered a very low form of

[1] This view is widely held by the Peasants of Rampura and elsewhere. See Section XI.

punishment]. Kenchayya was very annoyed with his mother over this, and some say that he started demanding the partition only after the above incident.

Kenchayya had demanded immediate partition on the previous night. His elder brothers were opposed to this—if a partition had to occur at all, they wanted it to occur only after another house had been built on the vacant site belonging to the joint family in the new extension in the village, and after the marriage of Kiri, the youngest brother, had been performed.

The *panchayatdars* or arbitrators regarded both these suggestions as reasonable. Sadhu's joint family was at that time living in a somewhat small house, built during Sadhu I's time, and if partition came into force at once, the house would have to be divided into four or five units by means of bamboo partitions. This was extremely inconvenient, and instead of making things easier, it would have made life more difficult for everyone. They would be getting in each other's way all the time, and quarrelling incessantly.

The arbitrators advised Kenchayya to wait till another house had been built, and Kenchayya agreed to the suggestioin, but insisted that it should be built within a year. To do this, the brothers would have to borrow money, an idea to which Sadhu II and Kempayya were both opposed. The latter said that given three or four years they would be able to build another house without having to borrow. [It is quite likely that they were trying to stall Kenchayya.]

Kiri's marriage too was an important matter. It is the responsibility of the head of a joint family to perform the marriages of all members who are of marriageable age. Kiri was 23 years old, and it was felt by all that his marriage was overdue. [Sadhu and his brothers had bought some land in 1945–6, and this probably came in the way of Kiri's marriage being performed. Among the Okkaligas of Rampura, marriage was much more expensive for the groom's people than it was for the bride's people. Kiri's marriage would have cost his joint family about Rs 1000 or more.[2]].

If the partition took place before Kiri's marriage, Kiri would have to remain with his mother or one of his brothers. Otherwise he would have no one to cook for him. It was also widely held that a young man who came into money would not be particularly wise or discreet, and marriage was regarded as an insurance against unwisdom and indiscretion. [The idea that a young man could not be trusted to spend his money wisely was consistent with the dependence of the younger members on the head of the joint family in all important matters.]

I have mentioned at the beginning of this section that both Uddaiah and Lakshmana visited me in the morning, and that, some-

---

[2]Kiri was allotted only Rs 700 for his wedding expenses in the partition deed, but this does not represent what the wedding would have actually cost his joint family had they performed it.

time after their arrival, Kempayya joined us. Kempayya listened to Lakshmana's version of the dispute. He only interrupted Lakshmana once to say that the arbitrators should have forced Kenchayya to remain united with his brothers. (Later Kempayya told me that the Headman and his sons were favouring Kenchayya at the expense of the others. This kind of feeling is common in a dispute, irrespective of the fact that the feeling may have no foundation in fact.) Uddaiah retorted that the three brothers could remain united if they liked and throw Kenchayya out. Kempayya thought over this remark and agreed that it could be done. (Incidentally, parties to the dispute constantly change their minds. This is either due to second thoughts replacing first ones, or due to the influence of friends. A few people are to be found in every village who take an interest in disputes *qua* disputes, do their utmost to bring about and promote them, and occasionally, even profit by them. Even those who have no such strong interest in disputes, see the humorous and entertaining aspects of a dispute and advice and information are freely offered.]

Kempayya left us, and Lakshmana told me that he did not agree with Kempayya that only Kenchayya wanted partition. According to him, everyone wanted partition, but only Kenchayya openly said so. Lakshmana stated further that Kempayya not only wanted partition, but wanted to become the guardian of the minor boy, Honnayya II. Guardianship has certain advantages: Honnayya's share of the land would be available at a nominal annual rent in grain; and besides, Honnayya's labour would also be available. Honnayya II was just then ten years of age, and a boy of ten is very useful—he can milk cows and buffaloes, graze them, plough land, and in addition, do a number of odd jobs about the house and farm.

Last night Kenchayya had proposed that Honnayya's share of the lands should be released to a tenant at eight *khandagas* (a *khandaga* = 180 seers) per annum. Eight *khandagas* are a high rent. He also proposed that the money obtained by selling the grain should be deposited with the Headman till Honnayya came of age. Kempayya rejected this suggestion. [Kenchayya's suggestion would have made him more popular with the Headman and his sons.]

As everyone had the feeling that Kenchayya was the one who wanted to break up the family, the Headman offered to take Kenchayya as his servant in place of Kempayya. Kempayya, as one of the servants of the Headman, spent most of the day, from about 6 a.m. till about 10 p.m., working for his master. At night he slept on the veranda of my house and not in his own house. If Kenchayya replaced Kempayya, he would be more or less effectively withdrawn from his brothers, though his wife, who was believed to be the ultimate source of the trouble, would be staying with them. [Kempayya was considered to be a shirker, and Kenchayya, a good workman, by the members of the Headman's joint family. Also Kempayya had the reputation of being too sly: two years ago he had successfully defied the Headman in the matter of the buying of a certain piece of land. The solution offered by the Headman it should be noted would have not only benefited Sadhu's family, but also that of the Headman.]

I have also the the impression that some of the moves made by Kenchayya were calculated to place himself in the good books of the Headman while doing damage to his brothers.

VI

The next entry about this partition in my notebooks is dated 15 March. I will summarize it here.

Kempayya seemed to think that his request that the partition be postponed by about two years was a very reasonable one. It would have enabled the joint family to clear the existing debts as well as build another house. Kempayya would have liked the Headman to force Kenchayya to remain united with his brothers, and he seemed disappointed that he did not do this. He also seemed to think that the Headman had agreed to Kenchayya's suggestion that Honnayya the minor should be allowed to choose his tenant. Kempayya feared that the boy would then go and live with his mother in her village, and that she would choose some stranger as a tenant, thus depriving one of the brothers of the tenancy. [Kempayya's fears proved to be without foundation: the Headman later insisted that Sadhu II should be the guardian of the minor as well as the tenant of his land.]

Kempayya was narrating his grievances to Uddaiah and Kulle Gowda—the latter was one of the cleverest men in the village and was helping me in the collection of information. He was also well-known for using every situation to his advantage. When Kempayya told him that he felt that the Headman ought to have been firmer with the recalcitrant Kenchayya, Kulle Gowda replied that the village doctor was at the bottom of all this trouble. The Doctor was an employee of the Mysore District Board, and he was liable to be transferred from one village to another. He was not a native of Rampura. Kulle Gowda and the Doctor had a quarrel when the village ceremonially observed mourning on the eleventh day of the assassination of Mahatma Gandhi [on 30 January 1948], and Kulle Gowda was anxious to get even with the Doctor. Kulle Gowda's attempt to make the Doctor the villain of the piece seemed very far-fetched to me but what surprised me was that Kempayya seemed to swallow it completely.

A few days later Kempayya informed that Kenchayya's wife was a relative of the Headman's mother. Kempayya seemed to imply that this was why the Headman was listening to Kenchayya. [This was, it is necessary to stress, only Kempayya's impression]. The Headman's mother was an elderly matriarch, held in great respect by the Headman and his offspring. It emerged from my subsequent inquiries that the Headman had arranged Kenchamma's marriage with Kenchayya, and that the girl was living for four or five years in his (Headman's) house before her marriage, as domestic help. It should be remembered that Kenchamma's inability to get on with her mother-in-law was immediately responsible for the partition.

VII

The next entry after 15 March is dated 2 May. The interval of six weeks is not easy to explain especially as a few important events did occur like the division of the lands, house, site, assets, and some movable property. The Headman and a few *panchayatdars* effected the division among the members of Sadhu's joint family. The terms of the division were, no doubt, going to be reduced to writing, but the Headman advised against the registration of the partition deed as that would mean expense, and make the terms of the deed inviolable. Mutual readjustments would then become impossible.

The brothers appeared dissatisfied with the way the land had been divided by the *panchayatdars*. They expressed their dissatisfaction to Kulle Gowda and suggested that he should redivide the land—a request which pleased him greatly as he was being called in to solve something which the Headman and other elders had not been able to solve. Sitting on my veranda he suggested a division on different lines. (Kulle Gowda, each of the brothers, and one or two others who were present, carried in their minds a picture of the lands the joint family owned.) Kulle Gowda's division seemed to meet with the approval of all, though it should not be surprising if they later discovered flaws in it. Kulle Gowda also advised registration of the deed. He expatiated on the disadvantages of non-registration. The deed would lay itself open to repudiation by any of the brothers subsequently. He quoted the example of Molle Gowda and brothers who had divided a year ago, but had failed to register the document. The brothers were now coming to blows on the fields, each wanting someone else's land.

The senior paternal uncle of Sadhu II who was present agreed with Kulle Gowda. He mentioned how when he and his brothers [including the late Sadhu I] partitioned their lands, they refused to register on the present Headman's father's advice. A year after the partition, however, the younger brothers said they wanted the shares of the elder brothers. As the latter were good people [he was paying a compliment to himself] they gave in to the demand of the younger brothers.

Kulle Gowda was happy that an uncle of the brothers had supported him. He assured the brothers that he would see that the registration went through smoothly. [Kulle Gowda was interested in the registration—it meant one or two trips to town for him. The brothers would have to bear the expenses of these trips, and besides, Kulle Gowda would be able to charge them other expenses, real or imaginary—the clerk had to be paid a certain sum for doing the work quickly not creating problems, and so on].

The incident narrated by the senior paternal uncle of Sadhu II brought out an important point viz., that when a dispute occurs, there are frequent references to precedents, which not only make the law clear, but perpetuate tradition and history.

The cattle, pots and pans, etc., had been divided a few nights previously under the supervision of Lakshmana and the senior paternal uncle. In none of the three partitions which occurred in the village during my stay was I able to witness the actual division of movables. This was probably due to the fact that the brothers who were dividing them wanted to hide their poverty from me, and also because they did not like a stranger to be a witness to their disagreements over the division of assets.

VIII

I have in my possession two partition deeds, one dated 8 May, and the other, 10 May. The former is incomplete, and it was abandoned in favour of the latter. A translation of only the second document is given below, though some important differences between the first and second are mentioned. Both of them were drawn up by Kulle Gowda, who might be called the chief scribe of Rampura, an occupation which gave him a little money, some power, and knowledge of the affairs of many, all three of which he valued highly. It is probable that in drawing up the document, Kulle Gowda had, as his model, a partition deed drawn up by a town lawyer. Kulle Gowda was very familiar with Mysore city, and he was more urbanized in his outlook than a great many in Rampura.

On this day, 10th May 1948, we the descendants of the late Sadhu I, viz., Sadhu II (about 40 years old), Kempayya (about 35), Honnayya II (about 10), son of the late Honnayya I, Kenchayya (about 31), and finally, Kriti (about 23), have partitioned our property, and the terms of the partition (*pālu pārikattu*) are given below. Before the *panchāyatdars* and before our mother Kempamma, we have divided equally our immovable property (*sthira āsthi*), and our assets and liabilities (*lēni dēni*). The shares of each one of us is separately liable for paying the assessment (*kandāya*) on his share of the land to the Government. We have earlier divided the movable property (*chara svattugalu*) before our mother and *panchāyatdars*2, and none of us has any complaints (*takarāru*) about the division.

As the fifth brother is as yet unmarried, Rs 700 have been given to him out of our assets for his wedding expenses. As Sadhu II's daughter has reached marriageable age, we give him Rs 150 for the expenses of her wedding. We give our mother Rs 218 (cash). As long as she is alive, each of us undertake to contribute to her maintenance 15 *kolagas* (a *kolaga* is a large paddy-measure equal to 9 seers) of paddy per year, measured in a nine-seer *kolaga* made in the Government Jail.[3] Each of us also undertakes to pay her Rs 5 per year. Besides, she is entitled to pluck as much as she needs of greens, lentils and vegetables in the lands of anyone of us. None may stop her.

'Sadhu II has been appointed guardian of the minor Honnayya II. The boy is quite fit (*lagattu*) for work. Sadhu II should make him work, and in return for his labour, give him food and clothing, and pay the assessment on his share of the land as well. In addition, Sadhu must pay Honnayya an annual rent of five *khandagas*[4] of paddy. This is the amount decided by the panchayat. Sadhu should sell the rent-paddy every year and deposit the money with the Headman. If any *panchayatdar* asks him at any time to produce the accounts, he should do so. If he does not, his share (*hissé*) of land will be held liable for the sum due.

'As our lands are adjacent, each has the right of passage on the lands of all the others, and also to receive water from the lands of all those above his own. (The country is undulating, and rice is cultivated in terraces at different levels.) The house in which we all live at present belongs henceforward only to Kempayya, Honnayya II and Kiri. To Sadhu II and Kenchayya we give the vacant site and Rs 1,000 (both to be equally divided between them.) As Kenchayya will have no house to stay in after the partition comes into force, we agree to allow him to stay for three months in our house during which time he is required to complete the building of his house. If, in future, any one of us, either out of his own volition or at the instigation of others, claims the share allotted to another, such a claim will not only be not conceded, but he will have to pay a fine of Rs 100. We have at present with us a pair of bullocks valued at Rs 275, cart valued at Rs 250, a sugarcane press valued at Rs 200, and the panchayat has agreed that Kempayya should have the press, and Kiri, the bullocks and cart. Consequently, Rs 200 shall be deducted from Kempayya's share of assets, and Rs 475 from Kiri's. Sadhu II, Honnayya's guardian, has taken over the assets as well as the liabilities of the minor.

We have earlier borrowed Rs 1,000 from the Headman. All the five of us are responsible for repaying the debt together with the interest due on the principal sum. Whoever fails to pay, his share of the land will be liable for the amount of the debt.

'Our twelve-pillared, tiled house in Rampura village in Hogur Hobli, Sangama Taluk, bounded on the north and south by streets, on the west

[3] The measures manufactured in the Government Jail were supposed to be acurate. Peasants were very particular about the measures used, as cheating during measuring was common, and assumed a variety of forms.

[4] A *Khandaga* was equal to 180 *seers* of paddy, and each *seer* equal to about 2½ lbs.

by— s house and on the east by—'s house, has been divided as follows: the area of five *ankanas* in the southern part of the house, from east to west, to Kempayya, an equal area in the northern part to Kiri, and the middle area to Honnayya II. The eastern veranda (*hasaru*) belongs jointly to Kiri and Honnayya, while the veranda-room in the west belongs to Kempayya'.

There were two schedules at the end of the document, one giving to assets, mostly loans, allotted to each, and the other, the amount of wet (*tari*) and dry (*khushki*) lands. They are combined into one schedule below:

Sr. No.	Name	Share of assets	Share of land			Remarks
			Wet	Dry	Total	
I	Sadhu II	Rs 844	1–7	0–24	1–31	
II	Kempayya	Rs 143	ditto	ditto	ditto	
III	Kenchayya	Rs 843	ditto	ditto	ditto	
IV	Kiri	Rs 569	1–9	ditto	1–33	
V	Honnayya II	Rs 246	1–7	ditto	1–31	
VI	Kempamma (mother)	Rs 218	—	—	—	

IX

There are some differences between the two versions of the document which call for comment.

In the first version Kempamma, Sadhu II's mother, was made the minor's guardian and Sadhu was given only the tenancy of the minor's land, though at a rent (five *khandagas*) which was much less than that which an outsider would have had to pay. Sadhu was asked to pay the rent to his mother who was made responsible for paying the assessment on the minor's land to the Government. She was required to hold the sums received in trust for the minor till he reached majority.

The guardianship of the minor was throughout a bone of contention. Sadhu's brothers seemed to have expressed a preference for their mother as a guardian, but the Headman firmly vetoed the idea and nominated Sadhu as a guardian instead. (It is likely that Kempamma was dissatisfied with the Headman's decision.)

Again, in the first version the ancestral house was divided between Kiri, Kenchayya and Honnayya II, while Kempayya and Sadhu II were asked to build houses on the vacant site. In the final version, on the other hand, Sadhu and Kenchayya were asked to move out of the ancestral house. This was because originally the younger men elected to stay in the house, but when it was decided that those going out of the ancestral house would be given Rs 500 each, to build a new house, Kenchayya decided to move out. When a partition takes place, a younger member has the choice of a share or benefit before an older member—the principle of seniority is reversed.

There are also differences in the details of drafting the deed which need not be gone into here. I will now comment on some of the features of the final deed.

The mother was given Rs 218 cash while the others were given loan documents. This was because of her sex and age—the realization of a loan was frequently a difficult matter. The mother was also given an amount of paddy which more than met her needs. Each sharer had to give her Rs 5 per year. Finally, she was given the right of collecting vegetables and lentils from the fields of all her children. (In another partition the Headman said, 'Women value such rights'). It was also decided that Kempamma was going to live with her eldest son after partition. Care was taken to see that the mother was looked after properly; an attitude of deep respect and regard prevailed towards her.

In the first version, Rs 750, and in the final, Rs 700, were set apart for Kiri's wedding expenses. It may be recalled here that many would have welcomed the partition to be postponed till after the wedding had been performed. Rs 150, were also allotted for the wedding of Sadhu's daughter. As already mentioned, among Okkaligas (Peasants) a son's marriage cost more money than a daughter's, and also, Kiri's marriage was a more direct charge on the joint family's funds than Sadhu's daughter's.

I regret to state that I did not inquire sufficiently deeply into the financial status of Sadhu's joint family. That is why I am unable to comprehend clearly the schedule at the end of the deed. Sadhu received Rs 844 presumably because the joint family owed him Rs 500 for quitting the ancestral house, and Rs 150 towards the expenses of his daughter's wedding. In addition, Sadhu took over the assets as well as the liabilities of Honnayya II.

I do not understand why Kenchayya was allotted Rs 843. It was true that he was also moving out of the ancestral house, but he did not have a daughter to marry off. Kempayya received the smallest amount as he was given the sugarcane press. Kiri seemed to benefit the most: besides the cart and bullocks, he received 2 extra *kuntas* (an acre is equal to 40 *kuntas*) of land,[5] and Rs 569. This was presumably

[5] It is likely these *kuntas* had been allotted to Sadhu II in the first instance, and Kiri bought them from him. Later, there arose a complaint that Kiri was refusing to pay for them.

because of the Rs 700 which the joint family owed him for his wedding expenses.

I am certain that the schedule had been drawn up after a thorough discussion of every item in it. There were some discrepancies, however, which I am unable to explain. But I am certain that none of the brothers would have quietly accepted what he regarded as an injustice.

X

11 May. Sadhu II called on me in the morning, and I had a talk with him about the partition. He complained to me that when the paddy in the granary was being divided among the brothers nobody mentioned that he should be given a little more than the others as he was the father of many children. (The expression *makkalondiga* or 'man with children' means that the man has many mouths to feed, and consequently was entitled to greater consideration from the others.) Sadhu II thought that he was entitled to receive more paddy than the other sharers.

Sadhu had complaints against both Kempayya and Kenchayya. 'Kempayya', he said, 'both pinches the baby as well as rocks the cradle'. He talked reasonably in public while covertly he caused a great deal of trouble.

Kenchayya had been keen on partition all long. He listened to his wife and took part in the women's quarrels; and while the partition was in progress he was throughout being advised by Uddaiah, one of the arbitrators. (Uddaiah seemed to be a sower of the seeds of discord.) Kenchayya caused a great deal of trouble by frequently changing his mind during the partition.

Sadhu II told me also the story of the late Honnayya I's efforts to enforce partition while their father Sadhu I was alive. Honnayya I demanded partition from his father only to be bluntly told that there would be no partition till all the sons had grown up and were married. He was also told that he was at liberty to leave the house and fend for himself. Honnayya left his paternal house in wrath, and joined his wife in her village, Hotte. Sometime later he fell seriously ill, and when he was dying, he sent for Sadhu II, his eldest brother, and not for his father. He requested Sadhu II, to look after his infant son Honnayya II, and to regard the boy as his own son. Sadhu II gave his word of honour that he would do so.

The boy continued to remain with his mother even after the death of Honnayya I. Sometime later, Honnayya's widow remarried, and Sadhu II thought of bringing the boy to Rampura. He thought of this partly because Kempayya, who had been married for several years, was without any children. [Subseqently, however, Kempayya's wife gave birth to a baby girl who was two years old at the time of the partition.] Both Kempayya and Kempamma objected to Sadhu's proposal. But, in spite of their opposition, Sadhu went to Hotte and brought the boy home. The boy's mother's relatives did not easily agree to surrendering him. But Sadhu was successful in the end, and for the last two years Honnayya was grazing the family's sheep and cattle.

As Sadhu was talking, the Headman came along and told him, 'I hope you will get Kiri married soon. Now that you have had your partition, you have no reason to quarrel. I hope you willk all unite . . . . It would have been really better if you had divided after another house had been built'. To which Sadhu replied that Kempayya was the chief obstacle to it. I was taken aback by this reply because until then Sadhu had been telling me that Kenchayya was the sole cause of the partition. [I had also been told by Lakshmana that it was Kenchayya who had wanted the partition to come into effect at once.]

12 May. Sadhu narrated an incident which had occurred in his house yesterday. He had killed a hen presented to him by his wife's people, and his wife had cooked it into a curry. As he sat down to his meal, he remembered that until recently he and all his brothers had sat down together to their meals. He then went and invited all his brothers to share the curry with him. Everyone came, barring Kenchayya, Sadhu is certain that Kenchamma prevented her husband from coming.

16 May. Kempayya, as has mentioned earlier, was *jita* (a traditional form of service prevalent in certain districts of Mysore State) servant in the Headman's house. There was an acute scarcity of land as well as labour in Rampura, and it was very likely that Sadhu II was asked by the Headman to persuade one of his brothers to be a servant in his houses, a request that would not have been easy to turn down. In return for his services, the Headman paid Kempayya a certain sum of money—frequently paid in advance—per year, and also gave him the tenancy of some land he was managing for a landowner who was resident in a town. The land of which the Headman had given tenancy to Kempayya was being jointly cultivated, before the partition, by the members of Sadhu's joint family. In one part of the land, sugarcane, and in another, rice, was being cultivated.

After the partition, Kenchayya claimed a share in the sugarcane crop as he had assisted Kempayya in its cultivation. But Kempayya replied, 'Either take over the tenancy completely, or leave me to do it.' Perhaps Kempayya said this easily as he knew that Kenchayya did not have a pair of bullocks with which to cultivate the land. Kenchayya took the matter to the Headman who supported Kempayya's stand. The Headman was interested in receiving the rent and any dispute between the brothers would have jeopardized it.

All the brothers deposited their respective loan-deeds with their mother, with the execption of Kenchayya who deposited his with the Headman.

On 20 May, the Headman went with two elders to the vacant site, which now belonged to Sadhu and Kenchayya, and divided it into two. On 21 May, the two brothers, their mother and the carpenter performed the ritual of 'worshipping the spade' (*guddali pūja*), which was performed before starting the construction of a house or hut. Kenchayya performed the menial tasks connected with the ritual, like digging a trench and watering it, while the mother brought the articles of worship on a tray, and Sadhu II actually performed the worship.

The disputes continued to occur. On 5 June, my diary had a brief: 'Kenchayya contnues to find grounds for disputing.' On 4 July, about seven weeks after the partition, Kenchayya moved from his ancestral house into the hut which he himself had built on the vacant site. A cock was killed on this occasion and a feast was given.

XI

2 July. Kempamma thought that the partition would mean ruin for her eldest son, Sadhu II, who had a big family. She also suspected that her dead husband's brothers were persuading Kiri to leave Sadhu and set up an independent house. Until then, on the advice of the Headman, Kiri was staying with Sadhu, an arrangement advantageous to both. But Kiri was already making it clear that he wanted to leave Sadhu, for instance, that morning he had gone to his own share of the land for agricultural work instead of going where Sadhu had asked him to go.

Kempamma, witnessing the efforts of Kiri to break away from Sadhu, had started saying that she did not want to be dependent on the goodwill of her sons for her livelihood. She was now claiming a share—five *kuntas* of land from each share—in the ancestral property instead of the maintenance allowance granted to her in the deed. Such a share in land was usually given to a widow—mother in Rampura, and was called *ajjipālu* or 'grandmother's share'.

It was reported that Kempayya was involved in a boundary dispute with one of his brothers. Kempayya was said to be working for his own ends and setting up Kiri against Sadhu. [According to one report, Sadhu's uncles were suspected of trying to divide Kiri from Sadhu, and according to another, Kempayya was accused of the same. Suspicions flew in all directions when a dispute was on. It was also not unlikely that old scores were paid off during disputes. All in all, the atmosphere got vicious, and relatives and friends came under suspicion.]

On the night of Friday, 2 July, I went to Sadhu's house with Kulle Gowda, and the Potter, Seekla, to help settle a dispute between the brothers. I am afraid my record of the discussion which took place is not very clear, and I have had to interpret and rewrite this particular entry considerably.

After the partitioin, Kempayya, a full-time servant of the Headman, had the unenviable task of cultivating his own share of the ancestral property as well as the land of which he held the tenancy. The partition took place in May, and June, July and August were three months of very heavy work on rice land.

Kempayya requested Sadhu to look after his land. Sadhu agreed, but sometime later, he began to ask in what way he was going to benefit by looking after Kempayya's land. (Sadhu had his own worries—his mother was asking for a share, and for the guardianship of Honnayya; and Kiri was threatening to leave him.) Perhaps he asked Seekla to find out from Kempayya what the latter was willing to pay. Seekla reported to Kempayya what the latter was willing to pay. Seekla reported that Kempayya had said to him that Sadhu was welcome to take both the sugarcane and paddy crops growing on the land of which he held the tenancy in return for cultivating his (Kempayya's) own share. But Kempayya denied that he had said such a thing to Seekla. Finally, it was agreed that Sadhu should take both the sugarcane and paddy crops growing on the Headman's land and that he should pay Kempayya a rent of 3½ *khandagas* of paddy.

As Kiri had worked on Kempayya's land from the date of partition till 2 July, Kempayya agreed to pay him Rs 4. He did not, however, pay any money to Sadhu.

While the talks were going on, Kiri, who was sitting in a corner, suddenly burst out, 'for the last 20 days, I have been made to work like a donkey. My chest is aching badly'. He started weeping. This upset everyone. It was a severe criticism of Sadhu's treatment of his youngest brother. He at once offered to pay Rs 20 for the work Kiri had done so far.

Kempamma said that she was tired of seeing her sons quarrelling. She did not want to be dependent on them for her livelihood. She wanted a share of her own to be carved out of the share of each brother. The sons expressed their willingness to give her a share. The mother's next demand was, 'what about fuel for me?'. Kulle Gowda told the brothers that they should give her a tree. Kempayya declared 'as long as I am alive, I will give her fuel'. Kempamma replied 'that is not what I mean. I want to be independent.'

On the following morning, I told a few elders who had gathered on my veranda that I was impressed by the eagerness of Sadhu and his brothers to accommodate their mother, and that this contrasted with the refusal to see each other's point of view. One of Sadhu's paternal uncles who was there, told me that when Sadhu I and his brother divided, each of them promised to give a certain amount of paddy every year towards the maintenance of their widow-mother. But Sadhu I defaulted once, and a few months later his mother died. The then Headman, the present Headman's father, refused to allow Sadhu I to take part in the funeral ritual.

This incident was narrated to me to show the respect due to a mother. The late Headman, the present Headman's father, was a remarkable man, greatly loved and respected by many. His action in keeping the defaulting son out of the funeral rites was approved by the people though the punishment was on the severe side.

Disputes continued to occur till the middle of November when I left Rampura. I have an entry towards the end of August in which Kempamma was reported as complaining that Sadhu's wife was not treating the minor Honnayya properly, and consequently, she wanted to look after the boy herself.

Sometime later I learnt that Kiri was refusing to pay Sadhu for the three *kuntas* of land which he had received from him. It is probable that this refers to the extra amount of land which Kiri received during the partition, in which case it was two *kuntas* and not three. But if they were given to him during the partition, he should be owing the money to all the brothers and not to Sadhu exclusively. It is possible that their transaction took place after the partition but I did not come to know of it.

Kiri was not the only person accused of not fulfilling the obligations which he undertook. All the brothers complained against each other. On 29 October, the brothers and their mother went to the Headman to lay the whole story of broken promises before him. The guardianship of the minor also came up for discussion, and the Headman told everyone that Sadhu would be the guardian and not Kempamma.

I met Sadhu on the following day and he told me that he was gathering coconut fronds and bark for thatcing the house he was going to build on his part of the site. He also told me that his wife's brothers (*bhāvamiadās*) living in Shivaputra village had promised him some bamboos for the roof. He added that his wife's people were 'good people'.

XII

The joint family under discussion came into existence when the late Sadhu I separated from his brothers and it came to an end with the present partition. It would have come to an end earlier had Sadhu I yielded to Honnayya I's demand for partition. It should be repeated that Honnayya I could have enforced his right to a share by going to a court of law, but the point to note is not the existence of a legal right, but of a custom overriding the right under certain circumstances.

While Sadhu I refused to divide the estate on the ground that two sons were still unmarried, Sadhu II and the *panchayatdars* were not able to have the partition postponed till Kiri's marriage. The probable reason for this was that, apart from the differences between the personalities of the actors on the two occasions, the normal authority of the eldest brother or mother was much less than that of the father.

It was only where the eldest brother was very senior to the others, and was also a man of exceptional character, that he commanded the same moral authority as the father. While the father was able to refuse a demand for partition, the eldest brother was not.

Honnayya I's leaving the parental roof and going to his wife's village highlights a feature of the kinship system of the Okkaligas of Mysore. Marriages forge ties not only between the bride and the groom but also between their respective kin-groups. In a patrilocal community, the girl leaves her natal kin-group to join her conjugal kin-group sometime after marriage. Marriage weakened a girl's ties with her natal kin-group, making her a member of her husband's kin-group. But what was not clearly recognized was that marriage, to some extent, also weakened a man's ties with his natal kingroup. (An astute elder of Rampura once told me that when a marriage took place it was not only a girl who was lost to her parents, but to some extent, the boy also, to his parents.) A man's ties with his brothers were markedly juridical, they were a matter of rights and duties. Conflicts were common even after partition. A man's relations with his affinal kindred, on the other hand, were marked by exchange of gifts and feasts. Sadhu II was given a hen by his wife's people, and later, they promised him bamboos for his house. He said that his wife's people were 'good people'.

Conflict between the mother-in-law and daughter-in-law was ubiquitous. Residence in the same house, and the fact that in the first few years of marriage, the mother-in-law was in some respects a more important relative than even the husband, increased the tension considerably. Kempamma beat Kenchamma with a broom; and later she complained that Sadhu II's wife was ill-treating Honnayya II.

It was common for the people of Rampura to say that the women who came into a joint family on marriage were responsible for its splitting. There was a substantial amount of truth in this statement. The brothers had many interests and experiences in common, and fraternal solidarity was always stressed in speech, and held up as an ideal. But the women who came into the joint family were strangers to each other, and they found the loyalty of a man to his brothers, sisters, and parents, very irksome, to say the least. Usually, the woman who came into the joint family by marriage was interested in making her husband break away from his brothers. It was part of her struggle for autonomy not only from her husband's brothers but from her mother in-law as well.

Mother-in-law and daughter-in-law quarrelled, a woman and her husband's brother's wife quarrelled, and finally, woman and her husband's sister quarrelled. But this did not mean that relation between the brothers was always one of harmony. In fact, the conventional explanation that the women who came into the family ultimately broke it may be regarded as a convenient myth, the function of which was to protect another myth which was fraternal solidarity. There are great tensions between brothers—there is economic rivalry, the younger brothers are expected to be subordinate to the eldest in every trivial detail of daily life which irks them, and so on. As long as brothers stayed united, the social personalities of younger brothers did not develop fully. The younger brothers were husbands and fathers, and heads of elementary families. But as members of the joint family they were subordinate to the eldest brother. The headship of an elementary family and membership of a joint family were in some respects incompatible. That was why the chances of a joint family breaking up became greater after the marriage of all the brothers. It was only where the joint family had a substantial amount of land or solid trading interests, or a great sense of family tradition, or when it was spread over a wide area that it continued to remain joint even after the brothers had become heads of elementary families. The management of huge properties required the co-operation of a number of people and this gave new opportunities to the joint family. The joint family also aided, if not necessitated, the expansion of family commerce: tensions between brothers or cousins diminished considerably when they did not share the same house.

The Potter and the Priest[1]

Characters in the dispute narrated below

1. Peasant KARAGU.
2. Peasant KEMPU, elder brother of 1.
3. Peasant MOLLE MARI, agnatic cousin of 1 and 2.
4. Lingayat BASAPPA.
5. Shephred CHIKKAVA alias JAVARAYI.
6. Potter NINGA.
7. Lingayat PUTTA, agnatic cousin of 4.
8. Oilman MADA.
9. Potter SUBBA, brother of 6 and seasonal labourer in the house of 7.
10. The VILLAGE ACCOUNTANT, a Brahmin.
11. Peasant JAPI, agnatic cousin of 1 and 2, and brother of 3.
12. Lingayat THAMMA, elder brother of 7 and head of the joint family.
13. Trader SAPPA has kept a grocery and cloth shop.
14. Peasant YANTRA, operator of the biggest rice mill in Rampura.
15. Peasant SWAMY, elder brother of 1 and 2.
16. Peasant NADU GOWDA, father of 1, 2 and 15.
17. Peasant HEADMAN of the village.
18. Peasant MILLAYYA, member of the same lineage as 15, and owner of he biggest rice mill in Rampura.
19. Peasant LAKSHMANA, 17's second son.
20. Lingayat MAHANT, a lawyer, elder brother of 4.
21. Lingayat SANNAPPA, elder brother of 20, and Food Depot Clerk in 18's mill.
22. Toddyman SENDI.
23. Peasant KULLE GOWDA, a busybody.
24. Lingayat KANNUR, a bachelor.
25. Peasant KARASI, a widow.
26. Peasant CHAMAYYA, CHIKKA DEVA, and CHELUVA, complainants.

[1]This paper was read at a seminar in the Department of Anthropology in the University of Chicago in the last week of May 1957. The award of a Fellowship by the Rockefeller Foundatioin for the academic year 1956–7 enabled me to spend the greater part of the year in working on my Rampura material. A full acknowledgement of the financial and other assistance received will be made later.

27. Peasant DADDA, son of KARI HONNU, agnatic kinsman of he Headman.
28. Smith SUBBA's wife

I

The dispute which I am about to describe occurred in the summer of 1952 when I was doing a second spell of fieldwork in Rampura, a village in the south-eastern part of Karnataka in South India. The earlier spell was ten months long in 1948, and the second trip was undertaken to cover certain gaps in the data. One of the subjects to which I wanted to pay some attention was the mode of settling disputes in the village.

I present the present case more or less it is in my notebooks as I want to convey an idea of how the dispute gradually unfolded itself to me. Each informant gave his version of the dispute, which not only added to the information which I had obtained from the others, but also modified and sometimes contracted it in some places. In doing so he revealed a new dimension to the dispute. It is hardly necessary for me to add that the version which emerges finally in this paper is only an approximation to the truth—closer probably to what actually took place than the earlier versions but still not *the* truth.

During my stay in Rampura in 1948, I was lucky enough to win the friendship of some of the influential young men in the village and when I returned in 1952, I told them that I wanted to understand how disputes were settled in the village and they agreed to be my mentors. In the course of hearing a dispute, one of my friends would turn to me and say, 'Come on, you give us your verdict'. This had the embarrassing effect of suddenly switching the attention from the disputants to me, and if I was foolhardy enough to accept the challenge and say something, my friends proceeded to dissect in public the implications of my verdict. They had no difficulty in showing what a great ignoramus I was, much to the amusement of the assembled crowd. This became a kind of side-show but I thought that I had to put up with it if I wanted to progress in my understanding of dispute settlement.

The main characters in this dispute are Ninga and Putta. The former was a Kumbara or Potter by caste but he did not pursue the traditional occupation of his caste. He owned very little land and he was the brother of Subba who was an agricultural servant in Putta's house. Putta was a Lingayat, a Shaivite non-Brahmin sectarian caste. Since the Lingayats had been recruited from a number of non-Brahmin castes, each Lingayat was found to be following the traditonal occupation of the caste to which he belonged before it came within the Lingayat fold. Some Lingayats were, however, priests. In Putta's agnatic lineage vested the hereditary priesthood of the temple of Mādeshwara in Gudi village which was only a mile away from Rampura. The priestly lineage lived in Rampura, and cultivated the lands with which the Madeshwara temple was endowed. The lineage was considered well off by village standards. The deity Madeshwara commanded a large following among villagers in this region.

II

I shall start the narration with an extract from my dairy,

> I went to Karagu's grocery shop. Karagu is an Okkaliga or Peasant by caste. I found sitting there, besides Karagu, his elder brother Kempu, their agnatic cousin Molle Mari, Lingayat Basappa, agnatic cousin of Putta, and Shepherd Chikkava alias Javarayi. They were engaged in an animated discussion. Kempu told me that there was an interesting dispute which he would describe to me and he would like me to give my verdict on it. Kempu gave me a brief account of the dispute.
>
> Potter Ninga had lent Rs 100 to Putta, younger brother of Thammayya. Ninga himself owed money to Oilman Mada and was about to clear it, when Putta took the money from him saying that he, Putta needed it urgently. Putta left for Mysore saying that he would be returning by the following morning, by the Lalita Bus. But Putta did not return as promised. The following evening Potter Ninga sat in a teashop abusing Putta. Shepherd Chikkava was also in the teashop at that time. The Potter said, 'May I sleep with the Priest's wife. May I sleep with his mother'. Then he said something even more serious. 'I am going to beat him with my sandals, and I am going to beat him till five pairs of sandals wear out'. [Leather defiles and if a man is beaten with sandals he loses his caste. He may be readmitted only after he has undergone an elaborate and expensive ritual or purification.]
>
> Chikkava was annoyed at the Potter abusing the Priest in so foul a manner. He got up to leave the teashop, and said, as he was leaving, 'Stop abusing. Remember that man is your guru'. [The Potter's and Priest's families had been friends for a long time, and at the time the dispute occurred, Ninga's elder brother Subba was working as a seasonal labourer in Putta's house. The Potters were in a sense the clients of the Priests. As the patron family were Priests, Putta was called Ninga's guru. While Subba and some others were clients of the Priests, Ninga was a kind of client of Peasant Kulle Gowda.]
>
> Someone went and informed Putta's brother and agnatic cousins about what had occurred in the teashop. The Priests then went in a body to the

teashop to beat up the Potter. But the Potter had managed to leave the teashop in the nick of time by a back door, and hid himself from the Priests for a while. Basappa, addressing Kempu, said, You Peasants will have to decide this case. We shall see how you settle it. We Lingayats are very few here'. [The Peasants were the dominant caste in the Rampura region. In Rampura they were little less than 50 per cent of the total strength of the village, and were wealthier than all the other castes put together. Disputes were referred to the Peasant elders by everyone and not merely by Peasants. The two most powerful Peasant elders were the village Headman or Patel, and Nadu Gowda, Kempu's father. It is relevant to point out in this connection that in the local hierarchy of castes, Lingayats were ritually higher than Peasants. But because the Peasants were numerically the largest caste, and were wealthy, everyone including Brahmins and Lingayats were secularly subordinate to them].

III

The case was then discussed by us. Karagu said that if the dispute was allowed to take a legalistic turn, it would make matters worse. He was for seeing the dispute in its proper perspective and for solving it soon. Otherwise it was likely to be blown up into a big thing. He also said that as the Priest was the richer party—'fat bottom' to quote him—, there was a tendency to take his side against the poor man. [Until then I had regarded Karagu as a placid person and I was surprised to find him so excited over this matter.] When Basappa said that even Shepherd Chikkava was angered by the abuse which the Potter had poured forth, Karagu countered, 'Perhaps Chikkava owes money to the Potter'.

Then Kempu playfully put the matter up to me for a decision. I rose to the bait. I said, 'It is a very simple case. It can be settled in a minute'

Kempu: 'Come on then. Give us your verdict'.

Srinivas: 'Both the parties are in the wrong here. The Priest failed to keep his promise while the Potter used grave abuse. The first should be fined Rs 5 and the second Rs 25.'

There was a chorus of approval. Molle Mari stated that this was exactly his decision the previous day.

I was called upon to state my reasons for such a verdict. While I conceded that the Potter had enough provocation, the mention of sandals was a serious matter indeed and he ought to be fined heavily. [Perhaps there would have been no case if 'sandals' had not been mentioned.]

Kempu, however, pointed out that a creditor has to be patient. He must expect the debtor to procrastinate repayment. For instance, he had lent Rs 50 to X over a year ago, and he had not been paid either principal or interest. Karagu answered his elder brother: 'The Potter is a poor man and he is not used to lending. Your arguments would have held good if he were wealthy.'

[In 1948, Basappa and his brothers had formally partitioned their property including the hereditary right to the priesthood of the Mādēshwara temple. During the course of partition relations had been strained between the brothers. But in the face of this attack by the Potter not only did Basappa unite with his brothers but the two branches of the lineage came together. In fact, it was even regarded as an insult to the entire caste of Lingayats in Rampura.]

Basappa said that had he met the Potter that day he would have beaten him throughly. Even now, when he thought of it, his stomach burned like a lime-kiln. He said, in his anger, that the matter would have to be referred to the elders of the Lingayat caste, and they would come and decide. Everyone deprecated this idea. The Lingayat elders would probably throw Putta out of caste. The Potter might reconcile himself to the loss of Rs 100 for the pleasure of seeing the Priest outcasted. Basappa replied, 'It does not matter. We shall take the matter to the caste elders'. It was obvious that he was highly incensed.

I asked Basappa, 'How can your caste elders have jurisdiction over the Potter? They can only bind you, but not the Potter. You will have to go to the Peasants in Rampura for justice'. Karagu, 'Don't allow this matter to become big. It is best to settle it here (in Rampura). If you press it too far, things may go against you'.

The Village Accountant chipped in, 'If you bring in outsiders, local people may refuse to give evidence'. One of the ways in which elders force the truth out of parties and witnesses is to ask them to swear in a temple as to the truth of their statement. Such a step is, however, serious, and the elders use it only as a last resort. But while they are able to put most villagers on oath, and elders of a caste can only put members of their own caste on oath. The others refuse to be put on oath.

I said, 'The matter could be closed by taking a small fine—say a rupee—from the Potter.'

Basappa: 'If the fine is so small, it will encourage everyone to indulge in abuse. The fine should be so heavy that he will remember it for the rest of his life'.

Basappa then cited the case of Kannur, a Lingayat bachelor, and Karasi, a young peasant widow. The two were discovered co-habiting, late one night, by a group of Peasant youths. They reported the matter to the village elders who promptly fined Kannur Rs 50. (The case is described at the end). Basappa seemed to think that the dispute was promptly settled, and Kannur was fined heavily because he was a Lingayat, and he had the audacity to sleep with a Peasant girl.

IV

Later that evening I ran into Japi, a young Peasant and a member of the same lineage as Kempu and Karagu, when I brought up the dispute in my talk with him. I told him that Karagu had been very sympathetic to the Potter and was against the Priest. Japi said that Karagu was ignorant of the real facts of the case. He then proceeded to give me his version of the dispute which revealed complications of which I was ignorant.

As I have mentioned earlier, Ninga's elder brother is an agricultural servant in the Priest's house. Subba's wages, amounting to about Rs 100, had not been paid. Subba was telling his close friends that he wanted to leave the service of Thammayya. When the latter was away in Ganjam village to buy a tree,[2] Subba requested Putta to pay his wages. Oilman Mada was demanding the repayment of the money he had lent Subba. Would Putta please pay? Putta gave Subba a hundred rupee note which Subba passed on to his younger brother Ninga to be returned to the creditor. Ninga went to Sappa's shop to get the note changed. As the note was being changed, Yantra, the operator of the rice mill, came into the shop, and requested Ninga for a loan of Rs 50. He said he would return the money, without fail, on the following morning. The shopkeeper told Yantra, 'If you are born to your father you will stick to your word'. Ninga lent him Rs 50, and then went to the teashop and had a large tea.

Very soon after he had paid Rs 100 to Subba, Putta came to learn that Subba was contemplating a change of masters. Putta realized that he had been tricked. He knew that his elder brother Thammayya would be very angry with him when he learnt about what had happened during his brief absence from Rampura. Putta decided to try and take back the money he had paid to Subba. He went to Subba and said he needed Rs 100 urgently. He would repay on the following morning. Together they went to the shopkeeper and recovered the note. [For the sake of greater clarity I should mention now alone that Subba had borrowed the money from Oilman Mada with his younger brother Ninga as his surety.]

The following morning I met Karagu and we talked about many things including the dispute. I told Karagu that I thought he was being partial to the Potter. Karagu replied, 'He is a poor man. Why should Putta have borrowed money from him? He is in the wrong to begin with'.

A villager who was present told Karagu, 'Ninga is weeping. You should see that he gets the money soon. Putta will be tempted not to return the money to the Potter. Ninga will be forced to complain to the Headman and Nadu Gowda.'

A few minutes later Ninga met me and I asked him, 'Why on earth did you drag in sandals?'. He seemed unrepentant. 'I shall say it again before the elders'.

[2] It had been decided to have a new Juggernaut made for the annual festival of the deity Madeshwara. Thammayya was busy touring neighbouring villages trying to buy a suitable tree for timber for the Juggernaut.

V

I must have had some doubts regarding the truth of Japi's version because I discussed it with Kempu and his elder brother Swamy. They said that Japi and Putta were friends, and Japi was naturally giving a version which favoured his friend.

I have now to make a brief digression.

In the summer of 1952, there was a serious split in the largest Peasant lineage in Rampura. Kempu's father Nadu Gowda was the traditional leader of this lineage. Nadu Gowda was wealthy by village standards, but the richest man in the lineage was, however, Millayya. In 1951 the Headman and Nadu Gowda both encouraged Millayya to start a mill to hull, clean and polish paddy, and a smaller one to grind rice flour. Millayya invested a considerable sum of money and started a big mill. A few weeks after the mill had started working, Kempu urged his father to install a mill of their own. Kempu was encouraged by the Headman's son, Lakshmana, in this enterprise. Millayya and his brothers were upset by this. They petitioned to the Government not to grant a licence for starting a second mill in Rampura. But Kempu, aided by Lakshmana, succeeded in securing a licence, and another mill was installed. It was smaller than Millaya's and it was made in Japan. Millayya's group carried on propaganda against Nadu Gowda's mill.

While all this was happening, Putta's agnatic cousin Mahant, a lawyer in a neighbouring town, had obtained a licence to install a paddy-huller. Sannappa, Mahant's elder brother, was at that time employed by the Government of Mysore as a Food Depot Clerk and he was friendly with Millayya. His office was in the mill itself, and he spent a great deal of his time in his office. He, Yantra, the mechanic of the mill, Japi and Putta were all working together in connection with the installation of a huller.[3]

Swamy and Kempu saw Millayya's hand in the huller. They believed that Millayya was encouraging the installation of the huller in order to take trade away from them. I remember Kempu telling me, in a different context, that he was annoyed that a fellow-casteman (Peasant) was injuring them and helping a Lingayat.

VI

Yet another version was forthcoming from Sendi: Subba urged Tham-

---

[3] In 1948 Japi installed a huller, with crude oil for fuel, in his house. He sold it, however, before Millayya and Nadu Gowda installed rice mills powered with electricity.

mayya to pay him his wages, and just before he left for Ganjam he told Putta, 'Give Subba four pallas of paddy'. (One palla is equal to 100 seers, and a seer is equal to a little over 2 lbs.) Putta replied, 'How can we pay him his wages when he still owes us three months of labour?'. Thammaya said, 'He will do it. Where will he go? He is with us. Give him the paddy.'

The Paddy was measured and given. Subba sold the paddy for Rs 108. Subba and Ninga together took the money to Oilman Mada's house. But Mada was not at home. As they came out of the house they met Yantra who asked for a loan of Rs 50. After some discussion, the brothers agreed to lend the money to Yantra. They went to Sappa's shop to get the note changed. Sappa did not have change for Rs 100. He had only Rs 50 which he gave Yantra. Putta owed Yantra Rs 25, and the latter requested the former for return of the loan before the following morning. Putta replied that he himself needed Rs 100 urgently. He requested Ninga to lend him Rs 100. He assured Ninga that he would secure the return of the loan-deed from Oilman Mada, and that no interest would be charged on the loan as from that day. Ninga handed the hundred-rupee note to Putta. Putta then left for Mysore without telling anyone. (This assumes that the hundred-rupee note remained with Ninga even after Sappa paid Rs 50 to Yantra.)

On the following morning, Oilman Mada asked Subba for the return of the loan. Subba took him to his younger brother. Ninga narrated to him what had happened on the previous day. Oilman Mada said Putta had not seen him at all. Ninga looked everywhere for Putta. It was only then that Ninga learnt that Putta had gone to Mysore the previous evening.

I told Sendi that I had already listened to three versions of the case and I wondered whether there was not a fourth. He said that his was the true version, and if he was proved to be lying he would pay me any damages that I cared to stipulate.

Both Swamy and Kempu were inclined to accept the latest version as true. They wondered whether the clerk Sannappa and his brothers had got together to get Ninga punished because Ninga was a close friend and follower of Kulle Gowda, who was the Food Depot Clerk before Sannappa, and from whom Sannappa had taken over. Kulle Gowda, angry at having to give way to Sannappa, had carried on propaganda against the latter. (I mentioned this here to illustrate how occasionally two big people fought each other through their clients or followers. Also the events which occur are given different interpretations in terms of previous relations among the various actors in the drama.)

Swamy said, ' I now see why Ninga abused Putta.'

I asked, ' Isn't there a tradition of friendship between the Priest's lineage and the Potter's?.'

Kempu said, 'There is. They will come together again'!

Kempu then said that the elders would abuse Basava. Poor man cannot escape abuse. They would also pull up Putta and make him return Rs 100.

I said, 'But Basappa seemed furious yesterday'.

Kempu said, 'They are furious before you and me. Do you think they will be furious before the Headman and Nadu Gowda?'.

VII

The Case of Kannur and Karasi, Daughter of Devi

It has already been mentioned that in the course of the discussion, Basappa referred to the case of Kannur and Karasi. Sometime before the occurrence of the dispute between the Potter and the Priest, Kannur, a Lingayat bachelor, was adjudged guilty of having had sexual relations with Karasi, a young Peasant widow. He was fined Rs 50, and she Rs 25, by the village elders.

I obtained a brief account of the case as it was quoted as a precedent. Since the events which are referred to below occurred several months previously, I do not know how accurate the account is. There were no documents against which I could have checked the version which I obtained. But I have no doubt, whatever, that it is at least broadly true.

Kannur was seen visiting Karasi's house by three Peasant youths, Chamayya, Chikka Deva and Cheluva. The complainants, it is alleged, were jealous as they had been snubbed by the girl in their efforts to get friendly with her. Kannur knew that the complainants wanted to catch him in *flagrante delicto*, and report him to the elders. He insulted the complainants saying. 'What can these Peasant youths do?' [He actually said, 'What can they pluck?' an expression which literally means that they are free to pluck his pubic hairs. The expression generally means that the speaker has utter contempt for his opponents.]

One night, sometime after midnight, Kannur was seen coming out of Karasi's house. Kempu said that Kannur was caught 30 yards away from the house. The complainants took him to the Headman. The latter sent for Swamy as his father Nadu Gowda was away on a pilgrimage of Banaras. But Swamy refused to go. Kannur was fined Rs 50 and Karasi Rs 25. Karasi was told that she ought to get remarried soon, and to someone in another village. Two or three days after the incident she left Rampura for her elder sister's village where she spent two or three months. She was then married in the *kuḍāvali* form (an abbreviated form for widows and divorcees). The Headman seemed to take a serious view of the incident. He told Kannur that he would be tied to a pillar in the Mari temple till the fine was paid. He was made to sit in the veranda of the temple. His fellow-casteman Thammayya paid the fine and freed him. Karasi had to leave her gold necklace with the Headman to be redeemed only after the fine had been paid.

Karasi's mother was known to be one of the masters of abuse in Rampura. This was a skill in which a few women excelled, and for which they were feared, disliked, and also admired, in the village. When the incident narrated above occurred, she was out of Rampura and was expected to return on the following morning. The complainants told the Headman, 'When Karasi's mother returns she will abuse us all. Karasi should see to it that we are not abused. If we are, she should be fined Rs 100.'

The mother returned on the following morning, and in spite of the warning issued by the Headman, she started abusing everyone involved in the previous night's incident. The complainant reported the matter to the Headman who, again, sent for all the elders in the village—Peasant, Millayya, Lakkayya, and Kempu (representing Nadu Gowda), and Lingayats, Thammayya and Sannappa. Rama, the Headman's eldest son was also present.

It was on this occasion that Kempu was irritated by the aggression and righteousness of the complainants, in particular that of Chikka Deva. Kempu knew that Chikka Deva's mother had slept with everyone including the Untouchables. He wanted to say, 'If there are holes in the pancakes everyone makes, there are holes in our pan'. (This remark has the same meaning as the Biblical one referring to the beam in one's own eye.) Ramu guessed that Kempu wanted to interfere on behalf of the respondents. So he winked to Kempu and Kempu took the hint and kept quiet.

Kempu boasted that he could have had the case against Kannur dismissed by asking the simple question, 'Was Kannur really caught red-handed?'.

I asked, 'Couldn't Kannur himself have asked that question?'

'No. They would then have made him take the oath'. Whereas if any arbitrator had raised the question, the case would have been dismissed for lack of evidence. One supposes that the friends of the respondent could have raised this point; but it was up to the arbitrators to decide whether to put a man on oath or not. A hostile arbitrator would not resist the temptation to use all the weapons at his disposal.

According to Basappa, on the second day, the Headman wanted to settle the case leniently and without much fuss as he had to take up another case in which an agnatic kinsman of his was involved as respondent. Dadda [son of Kari Honna] was accused of trying to sleep with Smith Subba's wife. The latter was walking near the Basava Pond when Dadda met her. It was not clear whether Dadda tried to rape her. Smith Subba's wife ran to the Headman and complained to him about Dadda's conduct. In his desire to see that Dadda was not punished severely, the Headman let off Karasi lightly.

Dadda's case was more serious than Kannur's as in the former, the husband was alive and Dadda's attentions were unwelcome. Finally, while in Kannur's case, the complainants were in no way relatead to either party, in Dadda's case, it was the victim herself who had complained. But the Headman merely abused Dadda and let it go at that.

VIII

General Remarks

Kempu and Swamy then said a few things about justice in village courts. Kempu stressed the distinction between a Government Law Court in the city and a court of village elders. The former is able to decide an issue entirely on its merits, while a village court has to look to the wealth and following of the disputants. If one of the disputants is capable of building a 'party' against the elders then the law is not strictly enforced. One has to allow 'the string to sag (*saḍila biḍabēku*)' for otherwise it will snap. Sometimes, facts have to be ignored. 'Let the facts slip through one's fingers' (*beraḷu sandiyalli biḍabēku*). An issue is sometime 'floated away' (*tēlisi bittēvu*) i.e. let off lightly.

One of the essentials in an elder is *jabardasti* i.e. a capacity to inspire fear in the disputants. If the parties are not afraid of the arbitrators, then there can be no justice. And one thing everyone was agreed upon was that the poor always got abuse from the arbitrators.

An example was given to show that where a man had a large following in the village he could defy the arbitrators. Eight or ten years prior to 1952, three close kinsmen of Nadu Gowda and Molle Gowda (father of Mari, Japi and Kempayya) stole paddy from a field just before the harvest. The culprits were fined heavily by the two elders. About the same time Shepherd Chenna, tenant of the Headman, stole horsegram leaves (fodder) from J.L. Sab's fields. Sab caught him red-handed and took him to the Headman. As Chenna was his favourite tenant, the Headman merely abused him and sent him away.

This made Molle Gowda very angry with the Headman, for, while he and Nadu Gowda had fined close kinsmen very heavily, the Headman had let go a mere tenant. He vowed to teach the Headman a lesson.

There was a village rule to the effect that during the paddy harvest, the shops which sell edibles, beedies and matches to the labourers were not to be stationed in the fields but inside the village site. This rule was enforced in order to prevent the theft of paddy. Molle Gowda asked his followers to put up shops everywhere in the fields. He did this to challenge the Headman to try and punish one of the offenders. The Headman realized that Molle Gowda's lineage, consisting of over thirty houses of Peasants, were with him in this matter, and kept quiet.

IX

Discussion of the Case

The dispute may be said to have a caste origin. While the Potter's expressing a wish to sleep with the Priest's wife and mother was certainly a serious abuse, threatening to beat him with sandals, was an even more serious one. While the abuse, 'May I sleep with your wife or mother?' is occasionally heard in rural areas, the other abuse, 'I shall beat you with sandals' is rare.

The seriousness of beating with sandals arose from the fact that leather defiled, and sandals came into contact with all kinds of dirt lying on the road. When Basappa threatened to take the matter to his caste elders, his friends warned him that it would probably result in Putta having to pay a fine and undergo the expensive ritual of purification. Putta, was, a priest and in an important temple at that, and the Potter's threat to beat him with sandals was likely to be interpreted by the caste elders as being equivalent to the act itself. I may mention here that a feature of village ethics which I had difficulty in comprehending was that when a case of sexual union between members of different castes was reported, the elder and more con-servative members thought as much of the resultant pollution as of the immorality of the act itself.

The dispute revealed the strength of caste as a principle of social affiliation, and it followed from this that village society was divided into as many layers as there were castes. The Peasant elders of Rampura of whom the Headman was, and still continues to be, the leader, acted with exemplary promptness in the case of Kannur and Karasi, and fined the former heavily, and made him stay in the veranda of the mari temple till the fine was paid. Basappa, a Lingayat, felt that all this was due to the fact that a Lingayat boy had slept with a Peasant girl. Basappa's agnatic cousin Thammayya, one of the village elders before whom Kannur and Karasi were tried, himself paid the fine because a man was expected to go to the aid of his casteman. Generally, the members of a caste living in a village were related by agnatic or affinal links. A casteman was also, frequently, a kinsman. He was also a kith.

The Peasant youths who apprehended Kannur were probably angry because a Peasant girl had chosen to confer her favours on a Lingayat youth in preference to themselves. Kannur expressed his contempt for the Peasant youths in such a way that it was also a challenge to their caste and to their manhood. This was probably why the Headman levied a heavy fine.

The harshness of the first day's verdict contrasted with the lenience of that of the second day's; Karasi's mother was not punished in spite of the threat of a heavy fine. Basappa attributed this to the fact that in the case which came up after Karasi's, the accused was an agnatic kinsman of the Headman. The kinsman was let off with abuse.

Kempu espied Millayya's hand in the Priest's attempt to obtain a licence for a paddy-huller and he interpreted it in caste terms. A more immediate idiom would have been the ties of lineage, and when I heard Kempu using the caste idiom, I thought it was due to Kempu's recent increased contact with towns. Lingayat–Peasant rivalry is a feature of urban Mysore, and Kempu was seeing village relations in urban terms. I mention this because nowadays urban influences reach the more accessible villages principally through youths who visit the towns frequently.

The Peasants were the dominant caste in Rampura—and also in the two paddy-growing districts of Mysore and Mandya—and some of the implications of their dominance were made clear in this dispute. The Lingayats were regarded as ritually superior to the Peasants, but as the latter were numerically preponderant and economically powerful, all the other castes including the Brahmins, were dependent upon them. This meant that the picture of caste hierarchy implicit in the all-India concept of varna did not have much meaning in the village, except in specified contexts. The economic and political power at the disposal of each of the castes varied from village to village and, therefore, each multi-caste village represented to some extent a unique instance of caste hierarchy. The dominance of a caste occasionally extended over a whole district or region, giving rise to a uniform hierarchical pattern over the entire area. But even in such a situation, there were differences from village to village.

Caste, however, was not the only principle of social affiliation. The village was another such affiliation, and it was based on the common interests which people inhabiting a restricted piece of territory posses-

THE POTTER AND THE PRIEST

LINGAYATS—PRIESTS

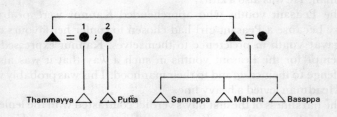

PEASANTS—OKKALIGAS

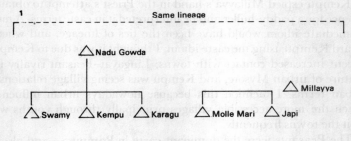

POTTERS—KUMBARAS

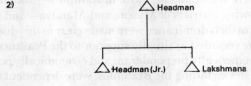

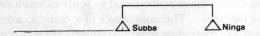

sed. Their common interests cut across caste. When Basappa mentioned that he wanted to take the dispute to the Lingayat elders, everyone was against the idea. The general opinion was in favour of the settlement of the dispute within the village. The dangers of taking it out of the village were pointed out. The court of village elders was a more friendly court, and more likely to take a lenient view of the case than a remote court of caste elders. What was even more significant, the elders of a caste would be helpless if they did not have the support and co-operation of the elders of the locally dominant caste. Only the latter were able to ensure the presence and co-operation of all the parties to a dispute—they had physical force at their disposal, they could enforce a boycott of the non-co-operator, and they could also bring economic sanctions to bear on him. The elders of any caste had to cultivate friendly relations with the elders of the village which in turn meant the elders of the locally dominant caste. Seen this way, caste and village were complementary and not conflicting but a clash between the two could occur.

Besides caste and village, the agnatic lineage, joint family and elementary family were the other elements of the social order. During 1948, Basappa was active in demanding partition from his brothers and on that occasion he made common cause with an agnatic cousin.

One explanation why the Priests wanted the case against Ninga to be treated seriously was that it afforded them an opportunity to wreck their vengeance against Ninga's friend and councellor, Kulle Gowda; Ninga had previously spent a few years as a farm labourer in Kulle Gowda's house. Occasionally a fight between two poor persons led to two more powerful persons entering the fight on either side. The poor persons then became pawns in a fight that was really between their powerful supporters. In the recent history of the village, it seems to have been frequent for patrons to fight each other through their respective clients.

Where a village was deeply factionalized, the overall authority of the elders was weak, and the leaders of each faction settled dispute between members of that faction. The village elders were not then able to enforce their decision on a faction which refused to submit itself to them. In such a situation village law resembled international law.

The ties binding individuals were multiplex and enduring. That was why Kempu was confident that the Potter and the Priest would come together again. Social relationships in rural areas were not specialized and temporary as in urban areas.

Village Studies, Participant Observation and Social Science Research in India[1]

I

I will not attempt to trace here the history of village studies in India but will rest content with mentioning that a few early British administrators did carry out, as part of their duties, village surveys in their respective areas. And during the first few decades of this century, a few foreign economists and anthropologists studied villages in different parts of the country, notable among them being Harold Mann, Gilbert Slater and the Wisers. Mann and Slater were interested primarily in the village economy while the Wisers were more comprehensive. William H. Wiser, as is well known to anthropologists, is the author of a brief monograph, *Hindu Jajmani System* (1936) and this was preceded by the better-known, little classic, *Behind Mud Walls* (1930), which he wrote jointly with his wife, Charlotte. The method adopted by the Wisers in their study of Karimnagar was quite different from that of the economists. They spent years in the village, talked to the local inhabitants in Hindi, participated in their activities, and did their utmost to help the needy and alleviate the sufferings of fellow-villagers.[2] The quality of the information gathered by the Wisers was superior to anything collected before, and when this was put into a holistic framework, which William Wiser's anthropological training enabled him to do, the result was a memorable picture not only of Karimnagar but of village life in the sub-continent, the microcosm reflecting the macrocosm.

It is only fair to add here that in the inter-War years Indian scholars such as D.R. Gadgil and R.K. Mukherjee also studied

[1] I would like to thank my colleagues the late, Dr S. Seshaiah and Drs J. P. Singh, M.N. Panini and Gurushri Swamy, for criticisms of an earlier draft of this paper. But needless to say, the responsibility for the views expressed is solely mine.

[2] The Wisers were missionaries but the point which I wish to stress is that they collected information through the use of the method of participant observation.

aspects of rural life while C.N. Vakil and G.S. Ghurye encouraged their students to carry out field-studies of villages. In retrospect, it is clear that this was a milestone in the development of the social sciences in India as it symbolized the recognition of the need, on the part of scholars, to find out the facts of rural life for themselves. No longer were they going to rest content with what was contained in the censuses, and administrators' reports and surveys. It signified a cognitive turn-around among the new elite cut off by class as well as caste from the rural poor. One wonders whether Gandhi's stressing of the importance of living in the villages to understand the conditions and problems of villagers was instrumental in bringing about this change among the Indian elite.

The situation with regard to village studies underwent a radical change after the end of World War II, when Indian social anthropologists, trained abroad, and their foreign counterparts, began to make systematic studies of villages in different parts of the country. These studies were different from the earlier ones, except, as already mentioned, that of the Wisers in that they relied almost exclusively on the method of participant observation, and the presentation of the data was usually around a well-defined theme of theoretical or comparative interest, though the theme used only a part, usually a minuscule part, of the information collected. In other words, there was no hiatus between description and theory, and instead the two were fused into a whole. The point I am making is well illustrated in the essays presented in *Village India* (Mckim Mariott 1955) and *Aspects of Caste in India, Ceylon and North-West Pakistan* (E. R. Leach 1960). The lack of focus which characterizes village surveys (witness the 1961 Census monographs) and many other studies, may be contrasted, for instance, with Adrian Mayer's *Caste and Kinship in Central India* (1960), or Andre Beteille's *Caste, Class and Power* (1965).

That facts are frequently marshalled to support or controvert an argument is well known, but what is insidious is the existence of bias in the perception and interpretation of facts. Training in research methodology ought to make students more sensitive and honest observers, and in this context the importance of contrary instances cannot be overemphasized. But as things are, courses in research methodology usually concentrate on survey techniques, elementary statistics, questionnaires, scaling and the like. Further, the importance of honesty and of contrary instances become devalued when it is argued that since the social sciences are not value-free, the pretence of objectivity should be discarded, and social scientists should use their knowledge and skills to promote worthy causes which are usually interpreted as latent causes. It does not seem to be realized that this argument could also be used by rightists and reactionaries.

In India, the fieldwork tradition is strongest in social anthropology and is only slightly less marked in sociology, while it is relatively weak in the other social sciences. It is only during the last twenty years or so that political scientists and public administrators have taken to fieldwork, and one of the factors underlying this shift was the desire to find out the forces underlying voting behaviour. Another factor was the desire on the part of the Planning Commission and the departments of the central and state governments to find out how the considerable sums of money they were spending on rural development were being utilized, and in particular, how different sections of rural population were benefiting from the development programmes.

Psychologists have been, on the whole, allergic to the dirt and dust of rural India, while in economics, the most developed of the social sciences in the country, fieldwork is regarded as an activity which is unworthy of the most sophisticated minds. The general tendency among post-independence economists has been to rely on data collected through the various agencies of the Planning Commission, the Reserve Bank of India, the ministries of the central and state governments, the census organization, the National Sample Survey, the reports of the U.N. agencies, and finally, the innumerable surveys conducted by government commissions and committees, research organizations, and teams of scholars. While economists have been diligent in extracting every ounce of information from the welter of published and mimeographed data, and have exercised their considerable (and expensive) skills in interpreting them, the idea that they should themselves undertake fieldwork does not seem to have occurred to them. I am now speaking of Indian economists as a general category and not of individual exceptions. This is indeed perplexing as they seem, again as a class, anxious to end exploitation of the poor and the oppressed, and to bring about equality and abolish proverty. But these laudable aims have not created in them a desire to come into close human contact with the objects of their concern and sympathy. (The only windows they seem to have to the world of the poor are servants, domestic and non-domestic.) In fact, I have had the somewhat disconcerting experience of listening to senior economists expressing bewilderment at the idea that a social anthropologist spends a year or two doing fieldwork in a *single* village, tribe, slum, or urban ward. How could busy intellectuals afford to spend so much time among illiterate, ignorant and susperstitious villagers? Whatever data that was needed could be collected in short periods of time by young investigators, or presumably older ones who were not good enough to make a contribution to analysis. The net result of this view is the emergence of a helot class of full-time data-gatherers who enable model-builders, theoreticians and other Brahmins to put forward their interpretations of the manner in which the economy functions. In this connection, I cannot help but recall the words of a brilliant Indian economist, 'why should an economist-planner know how to distinguish between rice and wheat stalks?' Why indeed, in a country where 80 per cent of the people live in villages, and are engaged in agriculture!

The blockage which economists seem to have against fieldwork seems to be so complete that it prevents them from even reading the writings of anthropologists concerning the economic behaviour of peasants. For instance, F.G. Bailey's *Caste and the Economic Frontier* (1957), in which he studies how and when a peasant's land comes into the market, seems to be almost unknown among Indian economists. Even more surprising, Scarlett Epstein's two books, *Economic Development and Social Change in South India* (1962), and *South India Yesterday, Today and Tomorrow* (1973), describing her studies of the divergent paths which development took in Wangala and Dalena, two villages in Mandya District in Karnataka, are also ignored. The reasons for this prejudice are perhaps complex and deep, but one explanation might be their conviction that it is a waste of time to find out what happens in single villages. Facts have to cover a wide region if they are to be 'useful', by which is meant relevant for policy formulation and planning. A completely instrumentalist view of social science seems to be axiomatic among economists if not most social scientists in India. In fact the few who are not convinced about the validity of this view find themselves in such a hopeless minority that they prefer to maintain silence at seminars, conferences and workshops.

II

The lack of a fieldwork tradition in the social science (excluding social anthropology and sociology) has had adverse results on their growth and development. Most important, it has alienated them from grassroots reality and led to fanciful assumptions about the behaviour of ordinary people. It has resulted in a woeful ignorance of the complex interaction of economic, political and social forces at local levels.

The educated Indian elite commonly regard the peasant as ignorant, tradition-bound, and resistant to progress. His actions and motivations appear anything but rational to the elite: his agriculture is 'gamble, on the monsoon', his plough just scratches the earth, his bullocks are bags of skin and bones, he indulges in wasteful expenditure at weddings and funerals, he is hopelessly indebted to the moneylender, his caste prejudices come in the way of his changing his occupation, and he lacks the sense to take advantage of the many benefits offered by a benevolent government working through a plethora of institutions and specialists.

The above image is only a caricature of the peasant, all the more untrue because it contains a grain of truth. Anyone who has tried to understand it knows that peasant agriculture is a highly-skilled enterprise, representing a subtle and delicate adjustment to an environment achieved over the centuries, and taking note of soil, weather, climate and other ecological conditions. These cultural traditions cannot be regarded as irrational as they represent among other things the peasant's psychological and social insurance in an extremely hostile world. Rationality does not exist in a vaccum but in a cultural context, and human satisfactions are themselves frequently culturally determined. The elites are annoyed with the peasant for not making choices which *they* want *him* to make, but they seem to be ignorant of the fact that choices are linked to structural, economic and cultural factors. This applies with particular force to a highly stratified and culturally diverse society such as India.

As in every other country, the poor in India have been subjected to gross exploitation and this has been both facilitated and compounded by the caste system, but this should not make us blind to the fact that over the millennia they have evolved an outlook and strategies which have enabled them to survive a variety of predators such as tax collectors, landowners, village headmen, accountants, money-lenders, marauders and the like. The morals and manners preached by the landowning high castes belong to a world different from that of the poor peasant.

Even the peasant's alleged stupidity, his inability to grasp and follow the instructions of a self-confident elite, is frequently a cloak for his scepticism towards new ideas and nostrums, a scepticism that stems from a deep faith in his tried and trusted ways. If a new variety of seed or mode of cultivation fails, the man who pays the price, a price which occasionally may mean starvation, is not the extension officer. Sometimes, however, the peasant senses that the new boon, institution, invention or technique advocated by the official is likely to place more power in the hands of rich and powerful men and increase his dependence on them. (On the other side, I have heard a powerful headman dismissing the idea of building a new school building on the ground that it would only teach the poor to be arrogant. The same man instead wanted electric power for the village as it would enable an industry to be started. And he would make sure that *he* would start the industry.)

Another consequence of the lack of serious fieldwork tradition in the social sciences is the implicit and absurd view that people are like dough in the hands of the planners and the government. Phrases such as 'social engineering', 'planned change' and 'directed social change' further the illusion that the government is able to change the lives of citizens in any manner it wants. That people have resources of their own, physical, intellectual and moral, and that they can use them to their advantage, is not recognized by those in power. Such blindness is all the more surprising since in most parts of the country the benefits of development programmes have been collared by the bigger landowners from the locally dominant castes, leaving the weaker sections of the society more or less where they were before. The point which I wish to make is that the well-made plans of the government do not go the way they are expected to, because people are *not* pawns which can be moved about but have intelligence, resources and aims of their own. They manage either to find loopholes in the rules, or bypass them altogether, in order to achieve their own ends, and in the process, defeat the planners. The tendency to study development programmes from the point of view of the degree of success achieved appears for me to be simple-minded. It would be more fruitful and interesting to look at them as providing examples of the responses of human beings with certain resources, values and aims, to the policies and programmes of a powerful government acting through its officials who not only wish to achieve certain targets but have a culture of their own, a culture that is mediated through their different personalities. It is a complex, interactional situation and constitutes an ideal area for research by social anthropologists.

The divorce between field research and theory has had another grave consequence for the development of the social sciences. Unlike the natural sciences, the social sciences have a deep and organic relation with the social (including the political and economic) environment which encompasses them and in which they function. In this sense they are relativistic, and this has to be accepted wholeheartedly if they have to develop and become an integral part of the country's intellectual life. They cannot hope to do this if Indian scholars keep continually looking to models, concepts and even empirical experiences from Europe, USA, USSR or China. To state this is not to be chauvinistic or to support obscurantism but only to make the obvious point that knowledge regarding an alien society and culture becomes meaningful only when compared with another of which one has first-hand knowledge. Such comparison is more often than not implicit than explicit, and that is certainly true in social anthropology. Further, Indian social scientists ought to realize that by concentrating on the Indian situation *they are really enhancing the range of the social sciences.* At the present moment, social sciences are drawing too heavily on a small range of human experience, viz., the Western-industrial, and equating it with the global. (Built into that equation is an ethno-centric assumption on the part of most Westerners that all societies are travelling towards the ultimate goal of Western-industrial society.) Indian social scientists have a responsibility to resist such an equation, firstly in order to better understand their society, and secondly, to contribute to the greater universalization of their disciplines.

The division of labour between the theoretician-analyst and the fact-gatherer is so firmly rooted in our country that I doubt whether it can be eradicated. Apart from organizations which already exist in order to conduct all-India surveys, the tendency to undertake big surveys will increase as administrators and politicians, partly for political reasons and partly to demonstrate their own sophistication, demand that facts be collected in a wide variety of areas for taking or implementing 'policy decisions'. (It would be instructive to discover in how many cases the information gathered has been used for the formulation or implementation of policy.) And during the last thirty years a class of academic-entrepreneurs has emerged in response to the demand for data and studies. They are usually senior men, heads of departments, or institutions, who enjoy directing projects, appointing staff, distributing patronage and flying to Delhi and other places for meetings of committees and seminars. Their intellectual interests are continually shifting to new areas: for instance, Indian social scientists have discovered poverty to be a profitable area for research. Research in slums and villages is also becoming fashionable.

The low qualifications, poor emoluments, harsh working conditions, and finally a lack of knowledge of, if not sympathy with, the aims of survey, which usually characterize investigators, result in the collection of information of dubious value. (I am aware that there are differences between various surveys in this matter but such differences are of degree and not of kind.) There are all kinds of hazards in such surveys: do the investigators understand the questions clearly, assuming, of course, the questions are unambiguous?, can they explain them to respondents?, how many of the questions touch on areas which are sensitive, especially during a period of ever-increasing governmental intrusion into the lives of the citizens. How many questionnaires does an investigator have to complete each day, and so on? When an investigator is required to complete so many questionnaires per day, and the latter are long and contain tricky, difficult and sensitive questions, then evasive, inaccurate and wrong answers are unavoidable. Even worse is the tendency to fake responses to questionnaires. As a callow researcher in Bombay, I was shocked when I first learnt of the faking of responses to a questionnaire on the sex habits of white collar employees in firms. The way the investigator obtained responses evoked hilarious laughter, and since then I have had a healthy distrust of data gathered by investigators, a distrust which has not diminished with the years. I recently came across an instance of an investigator who had faked, sitting in his home, responses to some forty questionnaires. Faking, however, is not confined to investigators. It also occurs at higher levels where data is processed—'laundered' might be a more appropriate term. For instance, when a macro-study involves several projects and the results of one of them diverge from the others, the latter may be brought 'in line'. The computing assistants feel safer when the results are not too diverse.

The bigger the survey, the greater the likelihood, if not the quantum, of faking. But academic decorum requires that this is not mentioned let alone discussed. But anyone who bothers will find out that underneath the decorous surface there is a whole body of folklore about how investigators fake information and how their supervisors fake supervision. The supervisory and investigating staff act in collusion, thus defeating the survey itself.

My remarks on the unreliability of the data collected in a macro-survey and the role of the research-patron are not peculiar to developing countries. For instance, according to Andreski:

> The chief advantage of the mechanical application of routine techniques is that it permits a massive production of printed matter without much mental effort. .... A research boss does not have to bother himself with observing or thinking about what he sees. All he has to do is to raise the money and recruit the staff who will do the work. Another advantage is that no matter how careless or even dishonest the interviewers might have been, the tabulated figures do not tell the story of how they came into existence, and the more massive the table the more inscrutable they become (1972, p. 109).

The belief seems to be widely held that all this can be taken care of by providing for a percentage of error in the responses. While such caution is to be welcomed, I find it difficult to believe that the investigator's failure to understand some questions, or clearly translate them, the numerous questions to which responses have to be obtained, the reluctance of respondents, and finally faking, can all be tidily summed up as a tiny percentage.

Among the hazards of data collection must be mentioned the resistance of respondents to part with information which they consider to be against their interests. Trader and small-scale industrialists do not like to answer questions relating to investment, profits, sources of finance any more than farmers like to answer how much wheat, rice and *jowar* they produced, or the precise area under each crop. The imposition of 'levy' on those growing staple grains such as wheat, rice and *jowar* is a disincentive to providing accurate information as far as farmers are concerned. And the farmers' lobby is so powerful at the state level that chief ministers are not likely to expert themselves to secure accurate figures regarding the area under each staple crop. The price paid for the grain surrendered as 'levy' is lower than the rate it would fetch outside, and given this fact, they are not likely to give accurate figures regarding the amount of grain produced.[3]

It is widely recognized that food statistics in India are politicized,

[3] When a state, or part of it, experiences less rainfall than usual, an effort is made to get that region classified by the central government as a 'drought area'. To qualify for such a classification, food production has to be below certain levels, and state authorities co-operate to produce the desired result. Central assistance is then given to the state government to undertake public works in the drought area and provide employment to the rural folk. Contractors not unnaturally welcome droughts as they are a source of income for them.

and further that different data-collecting organizations such as the Ministry of Food, Agricultural Prices Commision, and the N.S.S., each turn up with their own estimates regarding the amount of staple foodgrains produced in the country.[4]

The more knowledgeable men in the government recognize that food statistics only indicate a trend, but laymen, while welcoming the trend, cannot help wondering how that fact in itself can be of much use for policy formulation and administrative action. What is crucial is to have reasonably precise idea of the increase in each staple foodgrain, and this is so vital that a serious effort must be made to break through the obstacles in gathering accurate data.

But generally discussions on improving the quality of the data collected result in suggestions on increasing the size of the sample and for making it stratified, for better training and supervision of invetigators, for pre-testing questionnaries, etc. But the possibility of using a number of micro-studies of villages as checks against data collected from macro-surveys does not seem to have occurred to anyone since those in charge of the big surveys do not have any idea of micro-studies.

III

The fieldwork done by a social anthropologist offers a sharp contrast to the approach, methods and techniques used in macro-surveys. In the first place, a crucial characteristic of the anthropologist's study is the absence of any separation between the fieldworker and the analyst. Intensive fieldwork is now so integral a part of social anthropology that one who has not had the experience of carrying out at least one such study, is regarded as lacking a basic qualification. The total impact of such an experience is lasting in the sense of influencing for good the anthropological approach to the study of social institutions. The basic concepts of social anthropology are social structure and culture, and they result in emphasizing the need to view particular phenomena in their social and cultural context to ascertain properly their meaning and function. This to some extent underlies the annoyance, if not despair, at the manner in which questions are asked in macro-surveys: social facts are treated as though they are pebbles

[4] The Ministry of Food and Agriculture arrives at its estimate of food production on the basis of crop-cutting experiments in different parts of the country while the N.S.S. bases its calculations on per capita food consumption in the sample households. The Food Ministry's estimates tend to be lower than the N.S.S.'s

which can be lifted from a heap, an assumption that is frequently unconscious but nonetheless disastrous in as much as they ignore the vital connections which exist between social facts.

The method of participant observation has evolved through the efforts of successive generations of anthropologists, though the first practioner of the method was Malinowski.[5] Under this method, the fieldworker spends a period of 18–24 months in a society different from his own, the time being broken up in two visits, uses the local language in his talks with the indigenes, participates in their daily, seasonal and other activities, and as far as possible, gathers his information in the course of such participation. Before setting out for the field, the anthropologist is expected to be familiar with the litera-ture available on the region he has selected, and to have studied the language spoken there.

The lone anthropologist who goes out to a people to do fieldwork has a vital stake in collecting accurate information on the areas he considers relevant. He is aware that his prefessional future depends on the success of his first field effort. Even when he has a narrow and well-defined theme he collects a considerable amount of information on other matters as he realizes only too well that different aspects of social life are inter-meshed in a small community with unsophistica-ted technology.

As a result of their efforts, anthropologists have built up, over generations, a vast corpus of systematic information about the culture and social life of a large number of diverse groups, primitive, and now peasant, and on the whole, the quantity and quality of information is better than that available for the historic periods of civilized societies. Some anthropologists even regard the information that has been collected about the wide range of cultures as the crowning glory of their discipline. This pride is justified, particularly when it is re-membered that the spread of Western civilization over the globe has been destructive of cultural diversity.

As stated earlier, throughout the development of social anthropol-ogy, there has been an intimate linkage between fieldwork and theory. Richer data have given rise to new ideas and theory while theoretical developments have resulted in a demand for more and better data. As a result:

[5] For a brief account of this method, and the nature of the task which the social anthropologist sets out for himself when he undertakes a field-study, see Evans-Pritchard's *Social Anthropology* (1951).

(i) anthropologists are now more willing than before to quantify information and use statistical techniques; and

(ii) the range as well as the depth of the data collected has increased in response to new theoretical orientations and interests.

There is much more interest nowadays in ecology, in native perceptions and ideas, ethno-science, symbolism and ritual, and in social change, development and modernization. Intensive case studies are used in the analysis of selected aspects of social life, and this method has the advantage of dramatizing the interrelationship between diverse individuals and groups, the frequent clashes between norms and practice, and the creative manner in which leaders use situations to achieve their ends and strike out new paths. Ideally, a monograph can be written about a single case study, and this might scandalize those to whom 'methodology' means essentially the application of statistical techniques.

IV

I shall now touch briefly on the consequence of participant observation which, I think, reveals a dimension of social understanding not widely known.

There are two sides to the anthropologist's fieldwork, one which is open and exoteric, and the other, subterranean and esoteric. The former comprises carrying out censuses and surveys, collecting genealogical data, getting questionnaires completed, and recording events as they happen—be they life-cycle ritual, calendarical festivals, agricultural and other activities, disputes, partitions, transactions, jokes, abuse folklore, etc. The esoteric part, however, is more a byproduct of the method employed by the anthropologist than a set objective. The anthropologist's long residence among the indigenes, his knowledge of the local language, his participation in their day-to-day activities and his dependence on local people for the basic essentials of life including company, all produce over a period of time a change in his outlook and attitudes. The sense of strangeness, if not bewilderment, with which the anthropologist beings his fieldwork gives way, at some point along the way, to understanding which may be real, or partly real and partly illusory. He suddenly perceives the sense and rationality underlying indigenous institutions and behaviour, a perception the obverse of which is the awareness that the culture into which he was born was one of a myriad available to mankind, and any assumptions which he may have made about its being the only right and proper one, were absurd. Anthropologists are known frequently to become advocates, if not evangelists, for the cultures they studied as against their natal cultures. Even when it is only a passing phenomenon, it is indicative of the effects of the use of the method of participant observation.

This radical shift in the anthropologist's stance is not the result of cogitation but the chance outcome of living among the indigenes and sharing their life and experience. The change in his position is similar to that which a regular user of an automobile undergoes when he becomes, even for a brief while, a pedestrian, or that of a doctor who has to suddenly change his role for that of a patient.

The anthropologist's transformation is facilitated by certain objective conditions which have been mentioned earlier and which are incidental to the method he uses. In addition, however, he must have the gift of empathy, the ability to place himself in the shoes of others, and to look at the world from their point of view. This means that he should go much farther than the collection of accurate data. Ideally, he should have the sensitivity of a novelist.

The feeling that he understands the people he is studying is a heady one indeed, but is not exactly the ideal condition in which to write about them. A certain amount of distance, emotional and intellectual, is essential to the anthropologist for translating his personal and subjective experience into universal terms. This is helped, if not made possible, by the necessity to describe his experience in the language and concerns of his discipline at that moment. Seminars, informal discussions with colleagues, the need to compare, explicitly or implicitly, the institutions of the people he has studied with those of others, and above all, the fact that while writing up his data the anthropologist inhabits a different universe from that which he inhabited when doing fieldwork, all help put some distance between him and the people he studied. Reflection on one's fieldwork makes it more meaningful in the sense that the implications of seemingly trivial words, actions and gestures, which one had missed in the field, are brought home vividly. That it also leads to feelings of deep regret at missed opportunities is the other side of the coin.

The anthropologist knows that the information he collects during the first few weeks of his stay—frequently representing more than the total time spent by other social scientists on their fieldwork—is not only fragmentary but erroneous. While this is inevitable, the introductory period can be put to good use for collecting information of routine nature on non-sensitive areas such as material culture, agricultural practices, and genealogical data. The serendipity effects of this activity are at least as important as the information gathered, for in the course of going about his work the anthropologist becomes acquainted with the villagers and they come to accept his presence everywhere as a normal phenomenon. Time spent in establishing friendly relations with the people who are the object of investigation, is never wasted. It is the only means to the later securing of reliable information on sensitive areas. I am here deliberately ignoring the richness of the human experience of such encounters.

It is less difficult to look at the world from the point of view of the indigenes when they are a homogeneous group than when they are heterogeneous. The matter is compounded when indigenous society is hierarchical as the interests and perspectives of different groups conflict. The structural position of an individual largely determines how he perceives situations, and social interaction between people from different castes and categories involves a clash as well as an accommodation of diverse perspectives. Add to this personality differences and one gets some idea of the kind and degree of intellectual and emotional flexibility and sensitivity that the anthropologist ought to have to be able to study interaction between individuals in a complex and hierarchical society. He must not only understand the meaning of the words spoken but be sensitive to the tone and pitch of voice used, the accompanying gestures and facial expressions, and finally view all this in the total situational context of the transaction. Two identical words or phrases can have different meanings depending upon the context and the manner in which they were uttered.

The fruit of successful fieldwork is a large body of notes, the weight and complexity of which only increase with the years. Anthropologists who, for one reason or other, do not publish at least some of the results of their fieldwork soon after leaving the field, live to rue their failure. As they become more active in teaching—and this is particularly true of anthropologists in developing countries—they find that they do not have the time to write up their notes. After a while, the anthropologist discovers that he needs at least a few months even to become reasonably familiar with his notes, and writing them up is not something he can do in short spells of time but only as a sustained and full-time activity. The burden of unwritten notes is one with which most anthropologists have to learn to live. They come in the way of pursuing new interests. The value of old notes only increases with time: they become historical material.

V

I shall now refer briefly to a few common errors and misconceptions about village studies. An obvious objection to the study of a single village is its inability to inform us about the country as a whole. In one sense, the truth of this statement is obvious but in another sense it is not true. It is of course absurd to try and generalize on rural India from the study of a single village but if it is remembered that, in spite of its bewildering diversity, there are certain broad regional and even national similarities in India, even that study can be productive of knowledge and insights which could be translated into hypotheses and leads to future research. It also gives some idea of the quality of village life.

The Indian village has never been an isolate, being always connected, however tenuously, with neighbouring villages and towns in a variety of ways. Since independence the linkage between a village and its vicinity has grown more intimate. The social anthropologist who studies a village today will have to take serious note of its linkage with the wider society. While the village offers a convenient locus for fieldwork the anthropologist who studies it cannot stop at its boundaries. He has to follow his people to the other villages with which they have contact, to the *tehsil* and district headquarters, and even the state capital. Network analyses are likely to prove indispensable for analysing the nature of the relationship which a village has with neighbouring villages and towns. Even in a single village, they may differ for different areas of behaviour such as kinship, economics, politics and religion. It may also differ from group to group.

It is argued that the method of participant observation, apart from being time-consuming and productive of an unmanageable quantity of notes, is useful only for studying small communities. This becomes a serious limitation when there is a need for social anthropologists to address themselves to problems at regional and state, if not national levels. A great asset thus becomes a crippling liability.

There is some substance in the above argument. I would like, however, to distinguish between the value of the method of participant observation as a system of apprenticeship and its importance as a source of accurate and copious data. As already mentioned, an anthropologist's field experience influences his approach to the analysis of social institutions and the interpretation of social data, and finally enables him to refer what he reads and hears to his field-study.

In other words, his field experience is crucial for his intellectual development irrespective of the data it generates. Only a small part of the data produced may be used by him and he may not be able to use the rest.

The method of participant observation need not be confined to the study of communities. It can be used in case studies at various levels, from the national to local. A strike, a communal riot, a development project gone awry, the problems of an MLA or MP etc. Obviously, the purpose of a case study is to reveal in depth the play, interrelation and dynamics of various forces, and it makes no claim whatever to be representative. But even then case studies of the same phenomenon (riot or strike) can reveal a good deal about it which, apart from their intellectual interest, have practical implications.

Nothing that I have written so far about participant observation is inconsistent with the researcher's having a clear idea of what to look for in the field or formulating a hypothesis for testing. But such a course can only be taken when some reliable information is already available about the region. In the absence of such information, a general survey is unavoidable, though a researcher who has a sound grasp of theory and is also a keen observer is able to write a monograph with theoretical implications on the basis of his field experience.

Clarity about the aims of the project is essential in undertaking research at the regional or even higher level. The danger of dissipating one's energies in the pursuit of a vague and ill-defined objective is much greater when the area is large and the facts are not only more numerous but relationships between them extremely complex. Such projects are also likely to be costlier.

In brief, the village studies carried out by anthropologists in the 1950s were quite different from similar studies carried out by the other social scientists, and helped to make it clear to everyone that the former's interests were not confined to the tribal world. Since then they have extended their interests further to include pockets of the urban and industrial world. It appears only logical for them to move now to regional and even higher level studies. If this is to be done effectively there is a need for recasting syllabuses in social anthropolgy (and sociology). Apart from giving students in these disciplines a good knowledge of statistics and research methodology, provision must be made for their studying at least one of the allied social sciences, economics, political science and history. This is perhaps best done at post-graduate levels. Without such groundwork, talk about interdisciplinary work, which is both fashionable and fund-attracting, is a waste of time.

The return to macro- and micro-studies: with some thought and planning they can be made to support and strengthen each other. Correlations and generalizations derived from macro-studies may be examined in depth in micro-studies to find out if they are the outcome of real interrelations or only the accidental juxtaposition of unrelated events. Similarly, hypotheses and leads suggested by micro-studies can be tested systematically over wider regions.

Indeed, it would be worth while considering whether a certain number of villages from different regions of India, selected on the basis of relevant criteria, ought to be studied by anthropologists, economists and political scientists on a continuing basis and it would be illuminating to canvass in them questionnaires used in macro-surveys. The responses of people living in the selected villages can be compared against information obtained in other villages, and the divergences between the two sets, if any, should provide a check against the kind of information collected in the 'general' villages. The organization of these micro-studies would present problems and they would have to be gone into with care. But one thing ought to be clear: they ought not to become part of the empires of the managers of macrosurveys.

REFERENCES

ANDRESKI, S., 1972, *Social Science as Sorcery* (Andre Deutch, London).

BAILEY, F.G., 1957, *Caste and the Economic Frontier* (Manchester University Press, Manchester).

EPSTEIN, T.S., 1962, *Economic Development and Social Change in South India* (Manchester University Press, Manchester).

————, 1973, *South India: Yesterday, Today and Tomorrow* (Macmillan & Co, London).

EVANS-PRITCHARD, E.E., 1951, *Social Anthropolgy* (Cohen & West, London).

LEACH, E.R. (ed), 1960, *Aspects of Caste in India, Ceylon and North-West Pakistan* (Cambridge University Press, Cambridge).

MARRIOTT, McKin (ed), 1955, *Village India* (University of Chicago Press, Chicago).

WISER, W. H. and c., 1963 (1930) *Behind Mud Walls* (University of California Press, Berkeley and Los Angeles).

WISER, W.H., 1936, *The Hindu Jajmani System* (Lucknow Publishing House, Lucknow).

Index

Andreski, S. 100, 191
Anjaria, J. J. 28n, 58
Ansubel, H. 59
Ashe, Geoffrey 27n, 57
Avineri, S. 24, 57–8

Baden—Powell, B. H. 29–30, 57
Bailey, F. G. 185, 198
Beteille, Andre 183
Brahmins:
    attitudes towards 9–10
    and dominance 8
    and inequality 10
Brebner, J. B. 59
Buchanan, Francis 37n

caste:
    assembly 136
    council 104, 107–8
    hierarchy 79–86
    ideologies 47
    and kinship 134
    moral sanctions of 134–5
    and occupation 61–3
    and politics 12–13
    as a principle of social affiliation
        178–80
    and village 25
    complementarity of 180
caste ties, spread of 133–4
communal ownership 22–3, 26, 28
Cornwallis, Lord 37n

Crane, R. I. 26n, 57

Davis, Kingsley 39, 41, 57
decentralization 9
Desai, A. R. 28
disputes:
    dramatic side of 116
    and dominant caste 103–4, 106,
        108–10, 112
    nature of 2–3
    and non-official panchayats 122
    private side of 116–7
    role of panchayats in 136–7
    vested interest in 119–20
division of labour 24
documents referring to part
        disputes 121
dominance:
    and Brahmins 8
    consequences of 11
    criteria of 4–5, 11–12, 97
    elements of 6, 9, 97, 99
    function of 6;
    ritual 10–11
dominant caste: 17, 21, 29, 33–4,
        44–6, 55, 77–8, 96, 100, 104,
        113–14, 179
    council of 34, 103, 114–15
    definition of 4, 12, 77, 97
    and disputes 103–4, 106, 108–10,
        112
    exploitation by 102

(dominant caste contd.)
and power structure 3, 7–8, 13,
    102–3
    and Sanskritization 7
    secular control of 105
    and social order 9
Dube, S. C. 11–14, 18
Dumont, L. 3–7, 9–10, 19–22, 25,
    29, 36, 39–40, 45–6, 57–9
Dutt, R. C. 26n, 35n, 58

East India Company 24
economic self-sufficiency 36–9, 55
economy, traditional 70–4
elections and Untouchables 8
Elephinstone, M. 22, 29
Epstein, T. Scarlett 185, 190
Evans — Pritchard, E. E. 121,
    192, 198

factionalism 11, 42, 45
factions: 7, 180
    and dominant caste 8–9, 12, 42,
    112–13
Fryer, John 84
Frykenberg, R. E. 30, 58

Gadgil, D. R. 26n, 31, 37, 38n, 58,
    182
Gallahar, Miriam 20n
Gandhi, M. K. 26–7, 58, 183
Gardner, Peter M. 11, 19
Ghurye, G. S. 84, 95, 183
Gough, Kathleen E. 46, 58, 75, 95,
    112n, 115

Healey, J. M. 31–2, 58
horizontal:
    solidarity 43–4
    tie 7
Hunt, E. M. 59

inequalities 10, 12, 21, 23, 28, 40,
    43–4, 57

inequality:
    and Brahmins 10
    and Non-Brahmins 10
irrigation, importance of 35–6

jajmani (adade) system 11, 46, 66–7,
    72–4
jita 50, 87, 90
joint family 147–8
joint villages 34

Karan, P. P. 38, 58
kinship and caste 134
Kumar, Dharma 26, 28, 41–2, 58

Leach, E. R. 14, 19, 183, 198
Lewis, Oscar 46, 58
Lugard, Lord 33

macro-surveys and micro-studies,
    complementarity of 18, 198
Mahalingam, T. V. 35n, 58
Maine, Henry S. 20, 22, 24–8, 30–1,
    36, 41, 56, 58
Malcolm 22
Malinowski, B. 14, 192
Mann, Harold 182
Manu 62
marriage, cross — cousin 134
Marriott, Mckim 14, 19, 58, 75, 95,
    183, 198
Marx, Karl 20, 22–4, 26–8, 30, 56, 58
Mayer, A. C. 6, 46, 58, 183
Metcalfe, Charles, 21–2
Miller, E. 43n, 58
Morgan, L. H. 27
Mukherjee, R. K. 182
Munro, Thomas 21–2

Nanavati, M. B. 28n, 58
Nehru, S. S. 33
Non-Brahmin movement 98–9
Non-Brahmins and inequality 10
nucleated villages 60

objectivity 184

occupation and caste 61–3

O' Malley L. S. S. 21, 30–1, 35, 38

Oommen, T. K. 13n, 19

Panini, M. N. 182

Park, R. L. 26n

participant observation: 14, 16–17,
183, 192–5, 198
criticism of 17

Partition, role of women in 165–6

patron-client relationship 6–8, 11
42–2, 56, 86–91, 102, 104, 112

Pocock, D. F. 3–4, 9, 19–20, 39–40,
45–6, 58

politics and caste 12–13

pollution, concept of 126, 130, 133

primitive communism 24, 27

Rao, N. Rama 5, 19, 33, 58

rationality 186

ritual dominance 10–11

Rivett – Carnae 38n

Ruskin 27

Sanskritization: 2, 17, 99–100
and dominant caste 7
models of 7

Sastry, K. A. Nilakanta 35, 58

self-sufficiency, economic 36–9, 55
village 22–3, 28, 56

Seshaiah, S. 182n

severalty villages 34

Shah, A. M. 29, 121

Shivaji 33

Shroff, R. G. 121

Singh, J. P. 182n

Slater, Gilbert 182

Spear, Percival 21, 32–3, 41, 44, 51,
58

Srinivas, M. N. 5, 12, 19, 21, 29,
35, 45, 47n, 48n, 50, 58–9, 77,
95, 97, 115

Stokes, E. 22, 59

subjectivity 17

survey method 15–16, 184–9

swadesi movement 27

Swamy Gurushri 182n

tax– farming 33–4

Thoreau 27

Thorner, Alice 23, 41–2, 59

Thorner, Daniel 22n, 23–4, 28n, 30,
33, 35–6, 41–2, 59

Thurston, E. 83, 95

Tinker, I. 26n

Tippu Sultan 37n

Tolstoy, L. 27

traditional economy 70–4

Untouchables and elections 8

Vakil, C. N. 183

village:
and autonomy 30, 32
and caste 25
complementarity of 180
as a community 3–4
community 39–40, 56
antiquity of 25
councils (traditional) 103–4,
107–8
as a little republic 21, 27–9
resistance to change 23
self-sufficiency 22–3, 28, 56
as a social entity 44–7
society, role of brute force in 5
solidarity 20–1, 40, 51–5, 92
as a structural unity 91–3
studies, misconceptions about
196–7

villagers and state, the relations
between 29–36, 55, 57

villages:
joint 34
nucleated 60
severalty 34

vote-banks 17, 91, 102

weekly markets, importance of 37–8

Wiser, Charlotte 182, 198

Wiser, William H. 182, 198

Wodeyar, Chikka Deva Raja 74